THE ILLUSTRATED FLORA OF ILLINOIS

The Illustrated Flora of Illinois

ROBERT H. MOHLENBROCK, General Editor

THE ILLUSTRATED FLORA OF ILLINOIS

FLOWERING PLANTS
flowering rush to rushes

SECOND EDITION

Robert H. Mohlenbrock

SOUTHERN ILLINOIS UNIVERSITY PRESS
Carbondale

Library of Congress Cataloging-in-Publication Data

Mohlenbrock, Robert H., 1931–
    Flowering plants : flowering rush to rushes / Robert H. Mohlenbrock. — 2nd ed.
        p. cm. — (The illustrated flora of Illinois)
    Includes bibliographical references (p.   ) and index.
    1. Botany—Illinois. 2. Plants—Illinois—Identification. 3. Plants—Illinois—
Pictorial works. 4. Phytogeography—Illinois. I. Title. II. Series.
QK157.M6141984 2006
584'.8209773—dc22
ISBN-13: 978-0-8093-2687-7 (cloth : alk. paper)
ISBN-10: 0-8093-2687-6 (cloth : alk. paper)   2005035480

This book is dedicated to
my wife, Beverly,
who has proved a constant
source of
help and encouragement

CONTENTS

viii / CONTENTS

ILLUSTRATIONS

PREFACE TO THE SECOND EDITION

Since the publication of the first edition of *Flowering Rush to Rushes* in 1970, twenty-one additional taxa of plants covered by this book have been discovered in Illinois. In addition, numerous nomenclatural changes have occurred for plants already known from the state, and a myriad of new county records have been added. This second edition of *Flowering Rush to Rushes* is intended to update the status for the plant families covered in this book.

New illustrations have been provided for all of the additions except *Tradescanta subaspera* var. *montana* and *Najas guadalupensis* ssp. *olivacea.* All previously unpublished illustrations have been prepared by Paul W. Nelson, except for *Sagittaria platyphylla* and *Juncus validus,* which were prepared by Mark W. Mohlenbrock; and *Heteranthera multiflora* and *Eichhornia crassipes,* which were prepared by Phyllis Bick and are reprinted from *Steyermark's Flora of Missouri,* published by the Missouri Department of Conservation and the Missouri Botanical Garden Press.

PREFACE

After having worked with various aspects of the Illinois Flora for over a decade, I came to the realization that not a great amount of information was known about all the Illinois Flora, and that which was known was of a nature which was virtually useless to the average person wanting to know about the plants of this state. Thus the idea was conceived to attempt something that had never before been accomplished for any one of the United States—a multivolumed flora of the state of Illinois, to cover every group of plants, from algae and fungi through flowering plants. In addition to an account with keys of every plant known to occur in Illinois, there would be provided illustrations showing the diagnostic characters of each species.

An advisory board was set up in 1964 to screen, criticize and make suggestions for each volume of The Illustrated Flora of Illinois during its preparation. The board is composed of taxonomists eminent in their area of specialty—Dr. Gerald W. Prescott, Michigan State University (algae), Dr. Constantine J. Alexopoulos, University of Texas (fungi), Dr. Aaron J. Sharp, University of Tennessee (bryophytes), Dr. Rolla M. Tryon, Jr., The Gray Herbarium (ferns), and Dr. Robert F. Thorne, Rancho Santa Ana Botanical Garden (flowering plants).

This author is editor of the series, and will prepare many of the volumes. Specialists in various groups will be asked to prepare the sections of their special interest.

There is no definite sequence for publication of The Illustrated Flora of Illinois. Rather, volumes will appear as they are completed.

THE ILLUSTRATED FLORA OF ILLINOIS

FLOWERING PLANTS
flowering rush to rushes

WISCONSIN

IOWA

INDIANA

MISSOURI

KENTUCKY

County Map of Illinois

JO DAVIESS | STEPHENSON | WINNEBAGO | BOONE | McHENRY | LAKE
CARROLL | OGLE | DeKALB | KANE | COOK
WHITESIDE | LEE | DuPAGE
KENDALL
ROCK ISLAND | HENRY | BUREAU | LaSALLE | WILL
MERCER | PUTNAM | GRUNDY
STARK | MARSHALL | KANKAKEE
WARREN | KNOX | PEORIA | WOODFORD | LIVINGSTON
HENDERSON | IROQUOIS
HANCOCK | McDONOUGH | FULTON | TAZEWELL | McLEAN | FORD
MASON | VERMILION
SCHUYLER | LOGAN | DeWITT | CHAMPAIGN
ADAMS | BROWN | CASS | MENARD | PIATT
MORGAN | SANGAMON | MACON | DOUGLAS | EDGAR
PIKE | SCOTT | MOULTRIE | COLES | CLARK
GREENE | CHRISTIAN | SHELBY
MACOUPIN | MONTGOMERY | CUMBERLAND
JERSEY | EFFINGHAM | JASPER | CRAWFORD
CALHOUN | FAYETTE
MADISON | BOND | CLAY | RICHLAND | LAWRENCE
CLINTON | MARION | WABASH
ST. CLAIR | WAYNE | EDWARDS
MONROE | WASHINGTON | JEFFERSON
RANDOLPH | PERRY | HAMILTON | WHITE
FRANKLIN
JACKSON | WILLIAMSON | SALINE | GALL-ATIN
UNION | JOHNSON | POPE | HARDIN
ALEXANDER | PULASKI | MASSAC

Introduction

This is the first of many volumes devoted to the flowering plants of Illinois. The first six of these will be concerned with that group of flowering plants referred to by botanists as monocotyledons, or monocots. Technically, these are plants which produce, on the germination of the seed, a single "seed-leaf," or cotyledon, as opposed to dicotyledons, or dicots, which produce two "seed-leaves" upon germination of the seed. More practically, monocots are usually recognized by their long, slender, grass-like leaves and their flower parts often in threes or sixes, although this is not always the case.

Monocots include many plants familiar to most of us, such as lilies, irises, orchids, rushes, grasses, and sedges. This first volume begins with flowering rushes, follows with arrowheads, pondweeds, duckweeds, and cat-tails, and ends with spiderworts and rushes. The second volume commences with lilies, follows with irises, and concludes with orchids. The third and fourth volumes will be devoted to grasses, while the fifth and sixth volumes will include the sedges.

The nomenclature followed in this volume has been arrived at after lengthy consultation of recent floras and monographs of the plants concerned. The sequence of taxa presented here essentially follows that proposed for the forthcoming Flora North America series. It is basically a compromise of the systems of Cronquist (1968) and Thorne (1968). Synonyms, with complete author citation, which have applied to species in the northeastern United States, are given under each species. A description, while not necessarily intended to be complete, covers the more important features of the species.

The common name, or names, is the one used locally in Illinois. The habitat designation is not always the habitat throughout the range of the species, but only for it in Illinois. The over-all range for each species is given from the northeastern to the northwestern extremities, south to the southwestern limit, then eastward to the southeastern limit. The range has been compiled from various sources, including examination of herbarium material. A general statement is given concerning

3

the range of each species in Illinois. Dot maps showing county distribution of each monocot in Illinois are provided. Each dot represents a voucher specimen deposited in some herbarium. There has been no attempt to locate each dot with reference to the actual locality within each county.

The distribution has been compiled from field study as well as herbarium study. Herbaria from which specimens have been studied are the Field Museum of Natural History, Eastern Illinois University, the Gray Herbarium of Harvard University, Illinois Natural History Survey, Illinois State Museum, Missouri Botanical Garden, New York Botanical Garden, Southern Illinois University, the United States National Herbarium, the University of Illinois, and Western Illinois University. In addition, a few private collections have been examined.

Each species is illustrated, showing the habit as well as some of the distinguishing features in detail. Miriam Wysong Meyer has prepared all of the illustrations except for the Lemnaceae; these were illustrated by Dr. Kenneth L. Weik.

Several persons have given invaluable assistance in this study. Dr. Robert F. Thorne of the Rancho Santa Ana Botanical Garden and Mr. Floyd Swink of the Morton Arboretum have read and commented on the entire manuscript. For courtesies extended in their respective herbaria, the author is indebted to Dr. Robert A. Evers, Illinois Natural History Survey, Dr. G. Neville Jones, University of Illinois, Dr. Glen S. Winterringer, Illinois State Museum, Dr. Arthur Cronquist, New York Botanical Garden, Dr. Jason Swallen, the United States National Herbarium, Dr. Lorin I. Nevling, the Gray Herbarium, Dr. Robert Henry, Western Illinois University, Dr. John Ebinger, Eastern Illinois University, and Drs. George B. Van Schaack and Hugh Cutler, the Missouri Botanical Garden.

Southern Illinois University provided time and space for the preparation of this work. The Graduate School and the Mississippi Valley Investigations and its director, the late Dr. Charles C. Colby, all of Southern Illinois University, furnished funds for the field work and the salary for the illustrator.

Several persons have contributed portions of the text. Dr. Paul Fore prepared all of *Najas* and *Zannichellia;* Dr. Robert A. DeFilipps, all of *Juncus;* and Dr. Kenneth L. Weik, all of the *Lemnaceae*. Most of *Sagittaria* was contributed by Dr. James W. Richardson, while part of *Potamogeton* was written by Mr. James E. Ellis.

MORPHOLOGY OF THE ILLINOIS MONOCOTS

There is less diversity of growth forms in the monocots than in the dicots, due primarily to the lack of a vascular cambium in monocots. As a result, nearly all monocots are herbaceous. The only exceptions to this in Illinois are a few species of *Smilax*, with somewhat woody stems, and *Yucca filamentosa* var. *smalliana* and *Polianthes virginica*, which have a more or less woody caudex. Spines are present in the woody species of *Smilax*.

Many monocots are true aquatics which either are free-floaters or are rooted under water. Terrestrial species tend to be erect, although all species of *Smilax*, except *S. ecirrata* and *S. illinoensis* and the two species of *Dioscorea*, are vines. In *Smilax*, tendrils are present to assist in climbing. In *Echinodorus tenellus* var. *parvulus* and *E. cordifolius*, the stems may be creeping or arching.

Illinois monocots range in size from the minute species of *Wolffia* to *Yucca filamentosa* var. *smalliana*, whose inflorescences may reach a height of three meters.

Most monocots are chlorophyll-bearing, although chlorophyll is absent in *Thismia americana*, *Hexalectris spicata*, and the four species of *Corallorhiza*. In *Spiranthes gracilis*, *S. lacera*, *S. tuberosa*, *Aplectrum hyemale*, and *Tipularia discolor*, green leaves are often absent at flowering time.

In *Yucca filamentosa* var. *smalliana*, the leaves are evergreen. Certain of the woody species of *Smilax* may retain their leaves through most of the winter. In *Aplectrum hyemale* and *Tipularia discolor*, the leaves are present during the winter, but disappear in late spring.

The plants of the Lemnaceae are considered to be thalloid, *i.e.*, not differentiated into roots, stems, or leaves. In the other families, true leaves are usually present. These may be long, slender, and grass-like, or the blade may be broad. In *Scheuchzeria*, some species of *Juncus*, *Muscari atlanticum*, and some of *Allium*, terete leaves are encountered. Except in *Trillium*, *Medeola*, *Smilax*, some aroids, and *Dioscorea*, the leaves are parallel-veined, often conspicuously so. In many monocots, the leaves are sheathing at the base. A peculiar situation exists in *Limnobium spongia* where several of the central cells are inflated with air. This keeps the leaves buoyed up in the water and accounts for the common name sponge plant. In the Iridaceae, the leaves are characteristically compressed. Leaf arrangement is frequently basal or alternate, rarely opposite (*Najas*) or whorled (*Trillium, Medeola, Isotria, Elodea,* species of *Lilium*). Stipules generally are absent. While most leaves are green on the under surface, they are strongly glaucous below in *Smilax glauca* and deep purple below in *Tipularia discolor*.

Dimorphic leaves occur in some aquatic genera, notably *Potamogeton* and *Sagittaria*.

The underground structures are often bulbs or corms, although rhizomes and fibrous roots are not uncommon. In *Spiranthes* and *Triphora*, the roots are thickened and resemble tubers. Sometimes the covering of the bulbs is intricately sculptured, as in *Allium* species, and is an important identifying character.

Flowers are borne either from the axils or terminally. They may be solitary or arranged in various types of inflorescences, including cymes, corymbs, umbels, spikes, racemes, panicles, or even heads (*Xyris, Sparganium*). In the Araceae, the flowers are embedded in a fleshy structure, the spadix. Bracts usually subtend the flowers. They frequently are inconspicuous, but in *Commelina* and most members of the Araceae, they are large and are called spathes.

The flowers are usually perfect. Genera with unisexual flowers are *Sagittaria* (usually), *Najas, Zannichellia, Veratrum* (sometimes), *Melanthium* (sometimes), *Stenanthium* (sometimes), *Chamaelirium, Asparagus, Smilax, Elodea, Vallisneria, Limnobium, Arum, Peltandra, Arisaema, Lemna, Spirodela, Wolffia, Wolffiella, Typha, Sparganium,* and *Dioscorea*. Of those listed above, *Chamaelirium, Smilax, Elodea, Vallisneria, Limnobium,* and *Dioscorea* are dioecious.

The perianth generally is not well differentiated into calyx and corolla, although these are differentiated in *Alisma, Sagittaria, Echinodorus, Butomus, Xyris, Commelina, Tradescantia,* and *Trillium*.

There is no perianth in *Najas, Ruppia, Zannichellia, Peltandra, Arisaema, Arum,* or the Lemnaceae, while in *Typha* and *Sparganium,* the perianth is reduced to bristles and scales, respectively. In *Potamogeton,* the perianth-like structures may be sepals, or they may be expanded stamen connectives.

The predominant number of perianth parts per flower is six. In *Maianthemum* and *Symplocarpus,* the perianth is reduced to four segments. The parts of the perianth may be free, united only at the base, or united into a tube. In *Narcissus* and *Hymenocallis*, a corona is present.

The flowers are mostly radially symmetrical (actinomorphic), although zygomorphic flowers occur in some genera. The zygomorphic nature of these flowers is due to the presence of a modified petal (lip) in the Orchidaceae, the reduction in size of one of the petals in *Commelina*, the reduction in size of one of the sepals in *Vallisneria*, or the enlargement of one of the sepals in *Xyris*.

The number of fertile stamens often is three or six per flower, but several exceptions occur. In the Orchidaceae, there is only one stamen, except in *Cypripedium* where there are two. Various stamen numbers occur in the Alismaceae—six or nine in *Alisma*, six to thirty in *Echinodorus*, and indefinite in *Sagittaria*. Another family with variable stamen number is the Hydrocharitaceae. In this family *Vallisneria* usually has two stamens, *Elodea* has three to nine, and *Limnobium* has six to twelve.

Maianthemum, with four stamens, is the only exception from six in the Liliaceae. *Symplocarpus* has four stamens, *Peltandra* has four or five, *Arisaema* has two to five, *Potamogeton* has four, *Ruppia* has two, *Sparganium* has five, and *Typha* has one to seven. In the Zannichelliaceae, Najadaceae, and Lemnaceae, the stamens are reduced to one.

Staminodia are present in *Commelina, Pontederia, Peltandra, Arum,* and *Xyris*.

Apocarpy (the presence of more than one free pistil per flower) occurs in the primitive families. There are two to four free pistils in *Zannichellia,* and four in *Potamogeton* and *Ruppia*. The pistils become more numerous in the Alismaceae, where there are ten to twenty-five in *Alisma,* up to forty in *Sagittaria,* and more than ten in *Echinodorus*. In *Triglochin,* with three to six pistils, *Scheuchzeria,* with three, and *Butomus,* with six, the pistils are essentially free, but they are barely coherent at the base. In all other families, there is but one pistil. Sixteen families always have superior ovaries, while five families (Orchidaceae, Burmanniaceae, Hydrocharitaceae, Iridaceae, Dioscoreaceae) always have inferior ovaries. In the Liliaceae both superior and inferior ovaries can be found.

The capsule, which is the most common fruit type, is found in the following families: Commelinaceae, Orchidaceae, Pontederiaceae (except *Pontederia*), Iridaceae, Dioscoreaceae, Xyridaceae, Juncaceae, and some Liliaceae.

The fruit in Smilacaceae, Araceae, *Limnobium,* and some Liliaceae is a berry. In Juncaginaceae and Butomaceae, the fruit is a cluster of follicles.

The achene is the fruit in the Alismaceae, Najadaceae, Zannichelliaceae, Sparganiaceae, and Typhaceae. In the Lemnaceae and the genus *Pontederia,* the fruit is an utricle. The fruit is drupe-like but dry in the Potamogetonaceae, fleshy in the Burmanniaceae, and coriaceous in *Elodea.*

HABITATS OF ILLINOIS MONOCOTS

This discussion of the habitats of monocots in Illinois excludes the grasses and sedges which are treated in other volumes of The Illustrated Flora.

That monocots are predominant in moist situations is evident by the long list of species found in the aquatic and rich woods habitats. With somewhat fewer monocots are the dry woods and the prairies. Monocots of fields are very few, while adventives are slightly more numerous.

The habitats discussed below are the aquatic situations, the rich woods, the dry woods, the prairies, and the fields. In addition, adventive species and rare species are described.

AQUATIC SITUATIONS   The range of aquatic habitats for monocots progresses from continuous standing water to seasonal standing water of ditches and swamps to infrequent inundations in areas marginal to rivers and streams, and along lakes and ponds. In extreme northeastern Illinois, bogs present still another aquatic habitat for several monocots.

Usually rather common in standing water in most areas of Illinois are species of *Najas, Potamogeton, Sagittaria,* and the duckweed genera *Spirodela, Wolffia,* and *Lemna.* Some monocots in standing water are more prevalent in the northern half of the state. These include *Zannichellia palustris,* most species of *Sparganium, Elodea canadensis* and *E. nuttallii, Vallisneria americana, Pontederia cordata,* and *Zosterella dubia.* In fact, many of these are completely absent from the southern counties. On the other hand, *Wolffiella floridana,* a few species of *Spirodela* and *Lemna, Limnobium spongia,* and *Elodea densa* are the only species of standing water which are known from the southern counties, but not the northern ones.

Moist, mostly open situations, such as wet meadows, marshes, ditches, and borders of swamps, are usually rich in monocot species. Species met with frequently throughout the state are *Alisma subcordatum, Sagittaria* spp., *Iris shrevei, Acorus calamus,* and *Typha latifolia.* Northern species with few or no stations in southern Illinois are *Triglochin maritimum, Symplocarpus foetidus, Xyris torta,* and several species of *Habenaria.* Plants which are predominantly or exclusively southern are *Hymenocallis occidentalis, Iris brevicaulis, I. fulva, Habenaria peramoena, Echinodorus berteroi* var. *lanceolatus,* and *E. cordifolius.*

All of the Illinois bogs are in the northern section of the state. The monocots found in these bogs are often showy. They include *Cypripedium calceolus* var. *parviflorum, C. reginae, Habenaria hyperborea, H. psycodes, Liparis loeselii, Tofieldia glutinosa,* and *Scheuchzeria palustris* var. *americana.*

RICH WOODLANDS Rich woodlands in Illinois are characterized by dense shade, ample moisture, and a large amount of leaf litter. The overwhelming majority of the monocots in rich woodlands flowers during the spring. Characteristic vernal-flowering species are *Erythronium albidum, E. americanum, Uvularia grandiflora, U. sessilifolia, Allium tricoccum, Polygonatum commutatum, Smilacina racemosa, Arisaema triphyllum, A. dracontium, Cypripedium calceolus* var. *pubescens, Orchis spectabilis, Liparis liliifolia,* and species of *Trillium* and *Commelina.* Of more restricted range among the spring-flowering monocots which inhabit rich woods are northern species such as *Polygonatum pubescens, Smilacina stellata, Maianthemum canadense,* and *Medeola virginiana,* and southern species such as *Iris cristata* and *Aplectrum hyemale.*

The summer- and fall-flowering monocots of rich woodlands, except for the statewide *Spiranthes cernua,* have restricted ranges in Illinois. Essentially northern or central species are *Veratrum woodii, Allium cernuum, Liparis loeselii, Malaxis unifolia,* and several species of *Habenaria.* Predominantly southern species include *Hymenocallis occidentalis, Isotria verticillata,* and species of *Dioscorea.*

DRY WOODLANDS Less than 30 species of monocots in Illinois (excluding grasses and sedges) are characteristically inhabitants of dry woods. One of these, *Zigadenus glaucus,* is associated with exposed limestone in the northern tip of the state (although it also occurs in calcareous, springy areas), while *Polianthes virginica* is mostly restricted to exposed sandstone. The remaining species are in dry woods which may or may not exhibit exposed rock. Of these, *Lilium philadelphicum* var. *andinum* is restricted to the northern half of Illinois. Species confined to the southern counties are more numerous. These include *Tradescantia subaspera, T. virginiana, Nothoscordum bivalve, Dioscorea quaternata, Spiranthes tuberosa, Hexalectris spicata,* and several species of *Juncus* and *Smilax.* Found throughout the state in dry woodlands are *Tradescantia ohiensis, Allium canadense,* and *Dioscorea villosa.*

FIELDS Few monocots (excluding grasses and sedges) grow normally in open fields. Some which do are adventives, such as *Ornithogalum umbellatum, Muscari botryoides, Asparagus officinalis,* and *Allium vineale.* Native monocots of the field are *Commelina erecta, Juncus tenuis, Allium canadense, Sisyrinchium albidum,* and *Smilax lasioneuron.*

THE RARER SPECIES Of the 248 species of monocots (excluding grasses and sedges) known from Illinois, 25 per cent (63 species) are restricted to three or fewer stations. Twelve of these species are adventive in Illinois. In this section are descriptive accounts of the rare native species.

The most remarkable flowering plant of all in Illinois is *Thismia americana* (Burmanniaceae). This colorless plant, attaining a height of less than four centimeters, was discovered growing in a low, moist field outside Chicago by Dr. Norma Pfeiffer and a group of students from the University of Chicago in August, 1912. The station was revisited in 1913. The report of this remarkable plant, whose nearest relative is in Tasmania and New Zealand, appeared in *The Botanical Gazette* (57:122–35.1914), and was based on these two visits. Strangely enough, this tiny plant has never been seen again anywhere. Attempts have been made to rediscover the Chicago station, particularly by Dr. Julian A. Steyermark, then of the Field Museum of Natural History, and Mr. Floyd Swink, of the Morton Arboretum, but none has been successful. The rapid growth of Chicago into outlying areas soon spread over the only site where *Thismia americana* has ever been found.

An equally interesting story concerns *Lilium superbum* in Illinois. Mohlenbrock (1962) has described its discovery in detail. For several years, many of these lilies were observed in a mesic valley in Lake Murphysboro State Park (Jackson County), but withered before producing the flowers. Vegetatively the plants appeared to be *Lilium superbum*, but positive identification could be made only from flowers. Two specimens were transplanted to a wildflower garden in Murphysboro, and thrice have flowered. The large, handsome flowers verify the identification as *Lilium superbum*. One specimen of this species, rare in the mid-continent, finally attained the flowering stage in Illinois under natural conditions on July 6, 1966. In July, 1967, a flowering specimen was taken from Lusk Creek Canyon (Pope County).

The area where *Lilium superbum* occurs at Lake Murphysboro is within fifty yards of two other species known nowhere else in Illinois. Eastward and up-slope from *Lilium superbum*, in much drier soil, have been found *Smilax herbacea* and *Carex styloflexa* by the author in 1961.

Fourteen orchids have been discovered three or fewer times in Illinois. George Vasey, who was an ardent collector in bogs of northern Illinois during the last half of the nineteenth century, collected a specimen of *Malaxis monophylla* which is now deposited in the herbarium of the Missouri Botanical Garden. Unfortunately the data for the specimen are skimpy, merely stating "N. ILL." It is speculated that the locality was in McHenry County, where Vasey did most of his Illinois collecting. It is almost certain that this northern species is now extinct in Illinois. Another monocot collected only by Vasey from a bog in McHenry County is the tiny *Sparganium minimum* (Sparganiaceae).

Continuing with the Orchidaceae, six species of *Habenaria* have been collected three times or less. *Habenaria orbiculata*, *H. dilatata*, and *H. blephariglottis* are each known from a single station. The collections of *Habenaria orbiculata* and *H. dilatata* were made by Vasey from Kane and McHenry counties, respectively, during the last century. Neither has been collected since in Illinois. *Habenaria blephariglottis* was collected apparently only once from deep grass of a swamp near Decatur in Macon County in 1900. Collected twice has been *Habenaria hookeri*, first from Cook County as early as 1870, and later from adjacent Lake County. Both collections were made from rich woodlands. *Habenaria psycodes* is known from three counties in extreme northern Illinois, while *H. ciliaris* has been collected from three stations in Cook County, the first at Calumet by Hill in 1864.

Another specimen buried in the vast collections at the Missouri Botanical Garden is *Corallorhiza trifida,* collected in St. Clair County from a wooded hillside by John Kellogg on May 10, 1897. This is primarily a northern species which rarely penetrates southward, and the Kellogg collection is the only one known from Illinois.

One of the most beautiful orchids is the lady's-slipper, *Cypripedium acaule,* known only from Cook County. It was apparently first found in woods north of Stony Island (*fide* Pepoon, 1927), but no specimens seemingly were preserved. On May 20, 1908, while exploring deep wooded ravines near Glencoe, W. B. Day rediscovered this magnificent lady's-slipper. Then in 1942, Pearsall discovered this species at Elk Grove, Des Plaines, growing in association with the equally rare *Habenaria ciliaris.*

A rare lady's-slipper hybrid between *Cypripedium candidum* and *C. calceolus* var. *parviflorum,* known as *C.* × *andrewsii,* has been found only once in Illinois. It was collected by Mr. Virginius Chase near Spring Bay in Woodford County.

Spiranthes lucida apparently has been found twice, first in 1844 by Mead, north of Cain (Hancock County), and a little later by Agnes Chase in 1897 southeast of Marley (Will County).

Three rare orchids are confined to southern Illinois. The crested coralroot (*Hexalectris spicata*) was first collected from a rocky, wooded hillside near Prairie du Rocher (Randolph County) by Evers in 1949. The following year it was found by Hatcher in a similar habitat at Jackson Hollow (Pope County). Voigt, Mohlenbrock, and Sanders discovered it at Fountain Bluff (Jackson County) in 1954. Mohlenbrock and Voigt also made the first discovery of the rare crane-fly orchid (*Tipularia discolor*) from a wooded ravine in Jackson Hollow (Pope County) in 1959. This species subsequently has been found at several adjacent localities in Pope, Massac, Union, Gallatin, Hardin, and Johnson counties. Schwegman discovered some 125 plants of the rare *Isotria verticillata* in a springy woods in Pope County in 1967.

The iridaceous genus *Sisyrinchium* has its share of rare species in Illinois. In moist, usually sandy prairies in Cook and Lake counties has been found *Sisyrinchium montanum.* There appear to be only four Illinois collections, all made between

1900 and 1908. Also very rare in moist sand prairies is *Sisy-rinchium mucronatum*, first gathered by Mead in 1842 from Hancock County and later by Ahles in 1951 from near San Jose in Mason County. *Sisyrinchium atlanticum* has been found but once, from a wet prairie in Kankakee County.

A recently recognized species is *Camassia angusta*, a plant confused with the much more common *C. scilloides*. The only county in Illinois from which the rarer wild hyacinth has been collected is Macon. The two species occupy similar habitats.

The nodding trillium, *Trillium cernuum*, has not been seen in Illinois since 1891 when Hill collected it at Chicago's Wolf Lake (Cook County). Another rare trillium is *Trillium cuneatum*. This species has been collected twice, by the author, from a wooded oak-hickory slope at Giant City State Park, near the Jackson-Union County line.

Related to *Trillium* is the genus *Medeola*. The single species, *M. virginiana*, has been found four times in two counties. Its discoverer in Illinois was Locy, who collected it in a rich, wooded ravine near West LaGrange (Cook County) in 1877. Twelve years later, while collecting near Evanston in Cook County, L. N. Johnson found *Medeola virginiana* for its second record from Illinois. The last collection from Cook County was made by Buhl in 1916 from woodlands near Edgebrook. Fuller made a startling discovery of this plant from mesic woods near Ottawa, in LaSalle County, in 1939.

One of the spiderworts, *Tradescantia bracteata*, was originally discovered in Illinois from Morgan County in 1869; it has subsequently been found in Jersey County in 1947 by Evers, and in Mason County.

Two members of the Liliaceae qualify as rare in Illinois. *Zigadenus glaucus*, the death camas, is confined to the extreme northern counties of JoDaviess and Kane. It was first discovered by Vasey near Elgin during the nineteenth century, then recollected there by Benke in 1928 and others since then. In JoDaviess County, the death camas was first collected along the Apple River by Pepoon and Moffatt, and later in the same area by Ahles. *Chamaelirium luteum*, on the other hand, is confined to the southern counties (Hardin, Massac, and Pope). Pepoon and Barrett first discovered it in Metropolis (Massac County) in a "low thicket" in 1932. Twenty years later, Ahles collected this species from a wooded hillside near Lamb (Hardin County). In 1965, Schwegman discovered nearly one hundred plants growing in masses along a stream at Massac Tower (Pope County).

In 1838, George Engelmann collected *Heteranthera limosa* from the margins of a slough in St. Clair County. Since that time, this tiny species has been found at least five more times in St. Clair County. Then, in 1963, Fore and Stookey discovered *Heteranthera limosa* in a pond margin in Hardin County and, in 1968, Schwegman found it at Horseshoe Lake in Alexander County.

Another small amphibious taxon is *Echinodorus tenellus* var. *parvulus*. Mead collected it from Cass and Mason counties in the mid-1800's, and Eggert found it in St. Clair County in 1892. It has not been seen since in Illinois.

The genus *Potamogeton* has more rare species than any other monocot genus in Illinois. The rarest of these are *P. praelongus* (Cook, Lake, and McHenry counties), *P. berchtoldii* (Cook County by Hill in 1875 and 1880), *P. robbinsii* (Lake County, first by Hill in 1898), *P. strictifolius* (Cook County by A. Chase in 1900 and 1901), *P. vaseyi* (McHenry County by Vasey and Grundy County by Rowlatt), *P. × hagstromii* (Cook County, first by A. Chase, later by Sherff), *P. epihydrus* (Fulton, Hancock, and Lake counties, but not since mid-1800), *P. friesii* (Cook and Lake counties), *P. richardsonii* (Lake, Cook, and McHenry counties), *P. pulcher* (Menard, Mason, and Jackson counties), *P. gramineus* (Cook and Wabash counties), *P. × spathulaeformis* (Wabash County), and *P. × rectifolius* (Cook county, first by Hill in 1901).

Another rare aquatic is *Najas marina*, found in 1964 in Lake County by Winterringer. *Najas gracillima* has been found in three southern Illinois counties.

In the Juncaceae, *Luzula acuminata* is known only from LaSalle and Ogle counties, *Juncus vaseyi* only from Cook, McHenry, and Winnebago counties, and *Juncus scirpoides* only from Cass, Lawrence, and Menard counties.

SEQUENCE OF MONOCOT FAMILIES

The sequence of families of monocots and their placement into orders in most floras follow the arrangement of Engler or modifications of it. Thus the traditional division of monocots in Illinois is into nine orders. Detailed study of monocots during the past few decades by botanists has given us a new perspective on the phylogenetic relationship of the monocots.

The sequence of monocots presented in this series essentially follows that which will be employed in the forthcoming Flora North America. It reflects the thinking of two of the leading phylogenists in the country, Arthur Cronquist and Robert F. Thorne, who have devoted many years to the study of family relationships.

So that comparison may be made between the traditional arrangement of monocots and the arrangement found in this work, the two arrangements are outlined below.

Traditional Arrangement of Illinois Monocots	Arrangement of Illinois Monocots in this Work
Pandanales	Alismales
Typhaceae	Butomaceae
Sparganiaceae	Alismaceae
Alismales	Hydrocharitaceae
Najadaceae	Zosterales
Potamogetonaceae	Scheuchzeriaceae
Juncaginaceae	Potamogetonaceae
Alismaceae	Ruppiaceae
Butomaceae	Zannichelliaceae
Hydrocharitales	Najadales
Hydrocharitaceae ·	Najadaceae
Graminales	Arales
Gramineae	Araceae
Cyperaceae	Lemnaceae
Arales	Typhales
Araceae	Sparganiaceae
Lemnaceae	Typhaceae
Xyridales	Commelinales
Xyridaceae	Xyridaceae
Commelinaceae	Commelinaceae
Pontederiaceae	Pontederiaceae
Liliales	Juncaceae
Juncaceae	Cyperaceae
Liliaceae	Poaceae
Iridales	Liliales
Dioscoreaceae	Liliaceae
Amaryllidaceae	Smilacaceae
Iridaceae	Dioscoreaceae
Orchidales	Iridaceae
Burmanniaceae	Burmanniaceae
Orchidaceae	Orchidales
	Orchidaceae

The Alismales is considered first in this work under the belief that the most primitive families possessed more than one free carpel and a petaloid perianth. The primitive condition of the ovary and fruit is responsible for placing the Butomaceae before the Alismaceae. The Hydrocharitaceae, with inferior ovaries, is considered to be advanced in the order.

The traditional view that the Typhaceae, with its sparse perianth and reduced stamens and ovules, is most primitive among the monocots is not substantiated by morphological evidence. To the contrary, it is thought that the reduced number of flower parts usually indicates an advanced condition.

The next orders, Zosterales and Najadales, are the climax of the plants with free carpels which began with the Alismales. It would seem logical for the Juncaginaceae to follow the Alismales, the major difference being the uniformity of the two series of the perianth in the Juncaginaceae. The climax condition is seen in the reduction of all flower parts and in the aquatic habitat. The recognition of the genera *Potamogeton, Zannichellia,* and *Najas* into separate families is backed up by the morphological studies of Miki (1937) and Uhl (1947).

The Arales is an order composed of two families in Illinois. The Sparganiaceae and Typhaceae, following in the Typhales, seemingly are merely wind-pollinated aroids.

According to the sequence followed in this work, the Commelinales, which follow, are composed of the Xyridaceae, Commelinaceae, Pontederiaceae, Juncaceae, Cyperaceae, and Poaceae. These last two families are so large that they are treated in separate volumes in this series.

The Liliales, as constituted here, is composed of Engler's Liliales and Iridales. The order represents the beginning of the development of a perianth whose members are mostly uniform (frequently petal-like) and generally fused together, at least at the base.

It is within the Liliaceae that the classification followed in this work differs markedly from the traditional viewpoint. Not only are the usual liliaceous genera included (except *Smilax*), but so too are the traditional genera assigned to the Amaryllidaceae. *Smilax* is removed from the Liliaceae and placed in the Smilacaceae on the basis of the vining habit and the unisexual (in Illinois) flowers.

The remaining families assigned to the Liliales—Dioscoreaceae, Iridaceae, Burmanniaceae—all possess inferior ovaries. A tendency toward zygomorphy is climaxed in the Orchidaceae, the sole representative of the Orchidales.

HOW TO IDENTIFY A MONOCOT

A key to the identification of the monocot families in Illinois begins the systematic section of this volume. By use of this key, the families may be determined. The reader should then proceed to the family, where a key to the genera of that family in Illinois is provided.

Once the genus is ascertained, the reader should turn to that genus and use the key provided to the species of that genus if more than one species occurs in Illinois. Of course, if the genus is recognized at sight, then the general keys should be by-passed.

The keys in this work are dichotomous—*i.e.*, with pairs of contrasting statements. Always begin by reading both members of the first pair of characters. By choosing that statement which best fits the specimen to be identified, the reader will be guided to the next proper pair of statements. Eventually, a name will be derived.

Key to the FAMILIES of Monocotyledons in Illinois

* THOSE ENTRIES MARKED WITH AN ASTERISK WILL BE FOUND IN SUB-
SEQUENT VOLUMES.

1. Plants climbing or twining (if erect, then usually with a few weak tendrils from the upper axils); leaves net-veined; flowers unisexual.
    2. Inflorescence umbellate; ovary superior; fruit a berry_____
    _____Smilacaceae *
    2. Inflorescence glomerulate or paniculate; ovary inferior; fruit a capsule_____Dioscoreaceae *
1. Plants erect or floating in water (tendrils never present); leaves mostly parallel-veined; flowers bisexual or unisexual.
    3. Plants with one or two whorls of leaves.
        4. Flowers radially symmetrical; ovary superior; stamens 6.
            5. Plants never over 50 cm tall; flowers usually borne singly _____Trillium and Medeola, in Liliaceae *
            5. Plants more than 50 cm tall; flowers usually more than one _____Lilium, in Liliaceae *
        4. Flowers bilaterally symmetrical; ovary inferior; stamen 1____
        _____Isotria, in Orchidaceae *
    3. Plants with leaves alternate, opposite, basal, or none, or in several whorls.
        6. Flowers crowded together on a spadix, often subtended by a spathe_____Araceae, p. 124
        6. Flowers not crowded on a spadix (in Ruppia, two flowers are borne on a spadix-like structure).
            7. Plants thalloid, floating in water_____Lemnaceae, p. 138
            7. Plants with roots, stems, and leaves, aquatic or terrestrial.
                8. Perianth absent, or reduced to very minute scales (lodicules) or bristles.
                    9. Each flower subtended by one or more sterile scales; plants generally not true aquatics.
                        10. Leaves 2-ranked; sheaths usually open; stems usually hollow, with solid nodes, often terete; anthers attached above the base____Poaceae *

10. Leaves 3-ranked (when present); sheaths closed; stems solid, with soft nodes, often 3-angled; anthers attached at the base_____
_____ **Cyperaceae** °
9. Flowers not subtended by individual scales; plants mostly aquatics.
11. Plants erect; inflorescence terminal, spicate, thick; leaves very long, linear, strap-shaped___
_____**Typhaceae,** p. 163
11. Plants not erect, free-floating or sometimes rooted in bottom mud; inflorescence axillary or terminal and slenderly spicate; leaves not as above.
12. Leaves alternate; stamens 2 or 4; inflorescence spicate and usually terminal, or with flowers borne 2 per spadix.
13. Stamens 4; flowers in a spike or head; fruit sessile, appearing as an achene upon drying_____
_____**Potamogetonaceae,** p. 67
13. Stamens 2; flowers on a short spadix, concealed within the leaf sheath; fruit stipitate, drupe-like_____
_____**Ruppiaceae,** p. 109
12. Leaves opposite; stamen 1; inflorescence not spicate, axillary.
14. Carpel one; fruit beakless_____
_____**Najadaceae,** p. 115
14. Carpels 2–4; fruit beaked_____
_____**Zannichelliaceae,** p. 111
8. Perianth present, composed of either calyx or corolla or both (plants with the perianth reduced to minute scales or bristles should be sought under the first 8).
15. Pistils simple, more than one, separate or slightly coherent at base.
16. Calyx and corolla differentiated (in color and texture).
17. Inflorescence umbellate; pistils 6, coherent at base; fruit a follicle_____
_____**Butomaceae,** p. 22
17. Inflorescence not umbellate; pistils 10 or more, free to base; fruit an achene_____

_____**Alismaceae,** p. 23

16. Calyx and corolla undifferentiated (*i.e.,* similar in color and texture)_____

_____**Juncaginaceae,** p. 61

15. Pistil one, compound.
    18. Ovary superior.
        19. Calyx and corolla differentiated (in color and texture).
            20. Flowers crowded together in a dense head; leaves basal_**Xyridaceae,** p. 168
            20. Flowers borne in cymes or umbels; leaves cauline_____
            _____**Commelinaceae,** p. 172
        19. Calyx and corolla undifferentiated (*i.e.,* similar in color and texture).
            21. Flowers unisexual.
                22. Leaves net-veined; flowers in umbels_____**Smilacaceae** °
                22. Leaves parallel-veined; flowers in globose clusters, racemes, or panicles.
                    23. Perianth small, greenish; flowers aggregated in dense globose clusters; stamens 5____
                    _____**Sparganiaceae,** p. 155
                    23. Perianth usually conspicuous, greenish, yellowish, white, or bronze-purple; flowers in racemes or panicles; stamens 6
                    _____**Liliaceae** °
            21. Flowers bisexual.
                24. Perianth scarious_____
                _____**Juncaceae,** p. 194
                24. Perianth petaloid.
                    25. Stamens 3_____
                    _____**Pontederiaceae,** p. 187
                    25. Stamens 6(or 4).
                        26. Stamens of different sizes _____
                        __**Pontederiaceae,** p. 187
                        26. Stamens all alike.
                            27. Leaves evergreen,

rigid; stems woody__

Yucca, in **Liliaceae** *

27. Leaves deciduous, mostly not rigid; stems herbaceous__

_____**Liliaceae** *

18. Ovary inferior.

    28. Plants growing in water.

        29. Leaves whorled_____

_____**Hydrocharitaceae,** p. 52

        29. Leaves basal, or cauline and alternate.

            30. Stamens 2, or 6–12, never 3; flowers unisexual; styles not petaloid____**Hydrocharitaceae,** p. 52

            30. Stamens 3; flowers bisexual; styles petaloid_____**Iridaceae** *

    28. Plants growing on land.

        31. Flowers bilaterally symmetrical; stamens 1 or 2_____**Orchidaceae** *

        31. Flowers radially symmetrical or nearly so; stamens 3 or 6.

            32. Stamens 3; styles sometimes petaloid_____**Iridaceae** *

            32. Stamens 6; styles not petaloid.

                33. Leaves reduced to scales; plants lacking chlorophyll, at most 4 cm tall_____

_____**Burmanniaceae** *

                33. Leaves blade-bearing; plants with chlorophyll, well over 4 cm tall_____**Liliaceae** *

Descriptions and Illustrations

Order Alismales

Herbaceous perennials of marshes, with usually basal leaves; flowers bisexual or unisexual, regular; perianth composed of two distinct series of three members each, at least the inner series petaloid; stamens 6–40, spirally arranged; ovaries 6 or more, free or coherent only at the base; fruit an achene or follicle.

The Order Alismales in Illinois, as considered here, is composed of the families Butomaceae, Alismaceae (Alismataceae by some authors), and Hydrocharitaceae. The group is thought now by many taxonomists to represent the most primitive of the monocotyledons, primarily because of the several, distinct ovaries and the more or less distinguishable series in the perianth. The Alismales seems to be paralleled in the dicotyledons by the Ranales.

KEY TO THE FAMILIES OF Alismales IN ILLINOIS

1. Ovaries 6–10 or more, superior.
    2. Ovaries 6, coherent near base; inner series of perianth pink; inflorescence umbellate_____Butomaceae
    2. Ovaries 10 or more, free to base; inner series of perianth mostly white; inflorescence racemose or paniculate_____Alismaceae
1. Ovary 1, inferior_____Hydrocharitaceae

BUTOMACEÆ–FLOWERING RUSH FAMILY

Although Mohlenbrock and Richardson (1967) have speculated that the Alismaceae is a more primitive family than the Butomaceae, I now believe that the reverse is probably more nearly correct. The fruit in the Alismaceae is considerably more specialized and therefore most likely more advanced than in the Butomaceae.

Only the following genus comprises this family.

1. *Butomus* L. – Flowering Rush

Flowers perfect; perianth composed of two distinct series of 3 members each; stamens 9; ovaries 6, slightly coherent at the base; follicles many-seeded; leaves sedge-like.

Only the following naturalized species occurs in Illinois.

1. **Butomus umbellatus** L. Sp. Pl. 372. 1753. *Fig. 1.*

Perennial from a stocky rhizome; leaves basal, to 1 m long, 5–10 mm broad, glabrous; scape usually 1 m long, bearing a terminal umbel; umbel many-flowered; bracts 3, broadly lanceolate, acute, purplish; flowers pinkish, at least 18 mm broad; pedicels arched-ascending, to 10 mm long; sepals 3, pinkish or greenish, persistent; petals 3, pinkish, persistent; stamens 9; follicles to 1 cm long, long-beaked; 2n = 26 (Whitaker, 1934), 28, 40 (Lohammar, 1931), 39 (Gadella & Kliphuis, 1963).

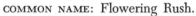

COMMON NAME: Flowering Rush.

HABITAT: Shallow water.

RANGE: Native to Europe; naturalized along the St. Lawrence River and around the Great Lakes.

ILLINOIS DISTRIBUTION: Known from a single locality in Cook County: near Buffalo Woods Forest Preserve, August 6, 1957, *F. A. Swink 3007.*

Illinois is one of several states where this species has been introduced in recent years. The pinkish perianth, nearly free ovaries, and many-seeded follicles readily distinguish this species.

ALISMACEÆ – WATER PLANTAIN FAMILY

Flowers radially symmetrical, perfect or unisexual; perianth composed of two distinct series, the inner white and petal-like; stamens 6–40; ovaries 10 or more, free from each other, each 1-celled; fruit an achene.

Within the Alismaceae, *Sagittaria* and *Echinodorus*, with their pistils borne in several rows on the convex receptacle, may be the primitive genera, while *Alisma*, with the pistils borne in a single row on the flat receptacle and with a more stable stamen number, may be advanced.

The strong morphological resemblance between *Sagittaria* and *Echinodorus* suggests an extremely close relationship. *Sagittaria* would seem to be primitive on the one hand because

1. Butomus umbellatus (Flowering Rush). *a.* Habit, X⅛. *b.* Fruiting branch, X¼. *c.* Flower, X1½. *d.* Cluster of fruits, X1½. *e.* Seed, X3.

of its large, indefinite number of stamens, and advanced on the other hand by its unisexual nature.

Care should be taken when collecting specimens of this family since both flowers and fruits are generally needed for positive identification. All members of the Alismaceae grow in or near water.

KEY TO THE GENERA OF Alismaceae IN ILLINOIS

1. Receptacle convex, bearing several rows of pistils; stamens 12–numerous (6–9 in *E. tenellus* var. *parvulus*); flowers unisexual or perfect.
    2. Achenes not winged; base of whorled inflorescence branches bearing 3 bracts and several bracteoles; flowers perfect; stamens never more than 21_____1. *Echinodorus*
    2. Achenes winged; base of whorled inflorescence branches bearing 3 bracts and 0 bracteoles; flowers mostly unisexual; stamens usually more than 21_____2. *Sagittaria*
1. Receptacle flat, bearing a single row of pistils; stamens 6–9; flowers perfect_____3. *Alisma*

1. *Echinodorus* RICH. – Burhead

Flowers perfect; receptacle convex; sepals 3, green, persistent; petals 3, white, deciduous; stamens 6–21; ovaries more than 10, distinct, arranged in more than one series on the receptacle; achenes turgid, ribbed throughout.

This genus is related closely to *Sagittaria*, but is separated by its wingless achenes, its bracteoles subtending the inflorescence branches, and its always perfect flowers.

Fassett (1955) has studied the tropical American species, which include our three representatives. The stamen number alone is sufficient to distinguish the three taxa in Illinois.

KEY TO THE TAXA OF Echinodorus IN ILLINOIS

1. Plants erect, less than 10 cm tall; leaves linear to lanceolate; flowers at most only 6 mm broad; stamens 6–9; achenes 10–15, beakless or nearly so_____1. *E. tenellus* var. *parvulus*
1. Plants erect, usually more than 10 cm tall, or plants creeping or arching; leaves broadly ovate, rarely lanceolate; flowers at least 8 mm broad; stamens 12–21; achenes more than 40, beaked.
    2. Scape erect; stamens 12; style longer than the ovary; beak of achene straight_____2. *E. berteroi* var. *lanceolatus*

2. Scape creeping or arching; stamens 21; style shorter than the ovary; beak of achene incurved_____3. *E. cordifolius*

1. **Echinodorus tenellus** (Mart.) Buchenau var. **parvulus** (Engelm.) Fassett, Rhodora 27:185. 1955. *Fig. 2.*

Alisma tenellum Mart. in Roem. & Schult. Syst. 7:1600. 1830.
Echinodorus parvulus Engelm. in Gray, Man. Bot. 438. 1856.
Echinodorus tenellus (Mart.) Buchenau, Abh. Naturw. Ver. Bremen 2:21. 1868.
Helianthium parvulum (Engelm.) Small, N. Am. Fl. 17:45. 1909.

Erect, rooted perennial, with creeping shoots frequently present; leaves basal, the blade linear to lanceolate, acute, tapering to base, to 3 cm long, glabrous, the petiole longer than the blade; scape to 10 cm tall, bearing a single whorl of 2–8 flowers; bracts 1–3 mm long; flowers at most only 6 mm broad; sepals suborbiculate, subacute, green, persistent in fruit, 1–2 mm long; petals suborbiculate, subacute, white, deciduous, 1–3 mm long; stamens 6–9; styles shorter than the ovaries; achenes 1.0–1.5 mm long, ribbed, beakless or with a minute lateral beak 0.3 mm long, glabrous.

COMMON NAME: Small Burhead.

HABITAT: Wet ditches.

RANGE: Massachusetts to southern Missouri, south to Texas and Florida; Cuba; Mexico.

ILLINOIS DISTRIBUTION: Very rare; specimens have been seen from only Cass and St. Clair counties. There is a report (Patterson, 1876) from Mason County. Typical var. *tenellus* has longer anthers and leaves concave toward the tip; it is restricted to South America. The variety *parvulus* in Illinois is readily distinguished by its 6–9 stamens.

For an extensive discussion of this species, see Robinson (1903).

2. **Echinodorus berteroi** (Spreng.) Fassett var. **lanceolatus** (Wats. & Coult.) Fassett, Rhodora 57:144. 1955. *Fig. 3.*

Alisma rostratum Nutt. Trans. Am. Phil. Soc. 5:159. 1837.
Echinodorus rostratus (Nutt.) Engelm. in Gray, Man. 460. 1848.

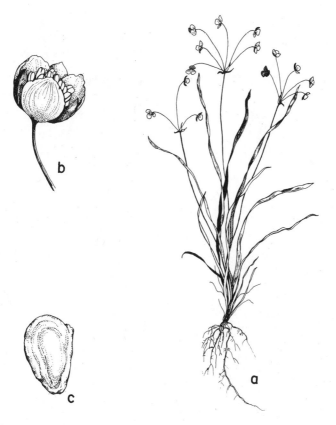

2. *Echinodorus tenellus* var. *parvulus* (Small Burhead). *a.* Habit, X½. *b.* Flower, X2½. *c.* Achene, X10.

Echinodorus rostratus var. *lanceolatus* Engelm. ex Wats. & Coult. in Gray, Man. 556. 1891.

Echinodorus cordifolius var. *lanceolatus* (Wats. & Coult.) Mack. & Bush, Man. Fl. Jackson Co. Mo. 10. 1902.

Echinodorus cordifolius f. *lanceolatus* (Wats. & Coult.) Fern. Rhodora 38:73. 1936.

Echinodorus rostratus f. *lanceolatus* (Wats. & Coult.) Fern. Rhodora 49:108. 1947.

Erect, rooted perennial; leaves basal, the blade broadly ovate, rarely lanceolate, obtuse, cordate or truncate at base, to 15 cm long, to 10 cm broad, glabrous, the petiole longer than the blade; scape usually more than 10 cm tall, with (1–) several

3. *Echinodorus berteroi* var. *lanceolatus* (Burhead). *a.* Habit, X⅙. *b.* Flower and buds, X1¼. *c.* Achene, X3¾.

clusters of 3–8 flowers; bracts 3–6 mm long; flowers 8–10 mm broad; sepals ovate, acute, green, persistent in fruit, 4–5 mm long; petals ovate to suborbiculate, acute, white, deciduous, 5–10 mm long; stamens 12; styles longer than the ovaries; achenes 2.5–3.5 mm long, with a straight beak about 1 mm long; 2n = 22 (Heiser & Whitaker, 1948).

COMMON NAME: Burhead.

HABITAT: Wet ditches; edges of swamps.

RANGE: Ohio to California, east to Texas and Florida; Delaware; Mexico.

ILLINOIS DISTRIBUTION: Local; scattered throughout the state; apparently absent from the northern one-fourth of Illinois, except for Carroll County.

The nomenclature for this taxon has been confused. For awhile, this species was erroneously called *E. cordifolius.* Then it was called *E. rostratus,* until

Fassett discovered the identity of *Alisma berteroi,* the basionym. Typical var. *berteroi,* from the southwestern United States, Central America, and the West Indies, has larger beaks on the nutlets and larger anthers.

3. **Echinodorus cordifolius** (L.) Griseb. Abh. Ges. Wiss. Gott. 7:257. 1857. *Fig. 4.*

Alisma cordifolia L. Sp. Pl. 343. 1753.

Sagittaria radicans Nutt. Trans. Am. Phil. Soc. 5:159. 1837.

Echinodorus radicans (Nutt.) Engelm. in Gray, Man. 460. 1848.

Creeping or arching perennial; leaves upright, broadly ovate, obtuse, cordate, to 20 cm long, to 17 cm broad, glabrous, conspicuously cross-veined on the lower surface; scape prostrate, creeping, over 50 cm long, bearing many whorls of flowers; bracts 10–25 mm long; flowers more than 12 mm broad; sepals ovate to suborbiculate, obtuse, green, persistent in fruit, 5–7 mm long; petals obovate, obtuse, white, deciduous, 6–12 mm long; stamens 21; styles shorter than the ovaries; achenes 1.8–2.2 mm long, ribbed, with an incurved beak about 0.5 mm long.

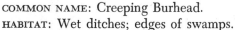

COMMON NAME: Creeping Burhead.

HABITAT: Wet ditches; edges of swamps.

RANGE: Maryland to Kansas, south to Texas and Florida; Mexico.

ILLINOIS DISTRIBUTION: Not common; restricted to the western side of the state, except for Crawford County. This taxon resembles *E. berteroi* var. *lanceolatus,* but differs by its arching growth habit, its 21 stamens, and the curved beak of its achenes. The creeping or arching stems further distinguish it.

During the period that *E. berteroi* (= *E. rostratus*) erroneously was called *E. cordifolius,* this species (the true *E. cordifolius*) was known as *E. radicans.*

2. Sagittaria L. – Arrowleaf

Mostly aquatic perennial herbs, with milky sap; leaves emersed, floating, or submersed, with sheathing petioles, the submersed leaves sometimes reduced to phyllodia; flowers whorled, bracteate, mostly unisexual, with the staminate flowers above the pistillate; sepals 3, greenish, persistent, those of the staminate

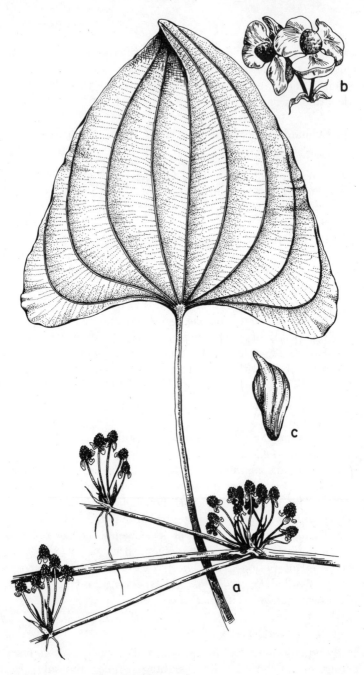

4. *Echinodorus cordifolius* (Creeping Burhead). *a.* Inflorescences and leaf, X⅜. *b.* Flowers, X½. *c.* Achene, X5.

flowers reflexed, those of the pistillate flowers appressed, spreading, or reflexed; petals 3, white or rarely pinkish, deciduous; stamens in a tight spiral, mostly numerous; carpels numerous, distinct, 1-celled, spirally arranged on a dome-shaped receptacle; achenes flat, winged, beaked. *Lophotocarpus* T. Dur. is included within *Sagittaria* since Bogin (1955) has pointed out that the usual distinguishing characteristics of *Lophotocarpus* of annuals with staminate flowers above and pistillate flowers below do not hold up.

Sagittaria is distinguished from *Echinodorus* by its flattened achenes and usual presence of separate staminate and pistillate flowers.

A recent revision of the genus is by Bogin (1955). Beal's (1964) treatment for this genus in the Carolinas seems to be most satisfactory when applied to Illinois material.

KEY TO THE TAXA OF Sagittaria IN ILLINOIS

1. Filaments roughened with minute scales.
    2. Peduncle and pedicel inflated; sepals tightly appressed to fruiting head_____1. S. calycina
    2. Peduncle and pedicel slender; sepals usually reflexed.
        3. Pistillate flowers and fruiting heads sessile or subsessile____ _____2. S. rigida
        3. Pistillate flowers and fruiting heads long-pedunculate_____ _____3. S. graminea
1. Filaments glabrous.
    4. Mature achene with 1–3 facial wings; beak of achene inserted apically.
        5. Beak of achene less than 0.5 mm long, straight, with 1 small, narrow facial wing_____4. S. cuneata
        5. Beak of achene more than 0.5 mm long, usually reflexed at apex, usually with 2–3 facial wings with 1 well developed.
            6. Receptacle not echinate; flowers in whorls of 3 or 4____ _____5. S. brevirostra
            6. Receptacle echinate; flowers in whorls of more than 4___ _____6. S. longirostra
    4. Mature achene with marginal wings only; beak of achene inserted laterally _____7. S. latifolia

1. **Sagittaria calycina** Engelm. in Torrey, Bot. Mex. Bound.
Surv. 212. 1859. *Fig. 5.*

Sagittaria calycina var. *maxima* Engelm. in Torrey, Bot.
Mex. Bound. Surv. 212. 1859.

Lophotocarpus calycinus (Engelm.) J. G. Sm. Mem. Torrey
Club 5:25. 1894.

Lophotocarpus depauperatus J. G. Sm. Ann. Rep. Missouri
Bot. Gard. 11:148. 1899.

Lophotocarpus calycinus f. *maximus* (Engelm.) Fern. Rhodora 38:73. 1936.

Sagittaria montevidensis ssp. *calycina* (Engelm.) Bogin,
Mem. N. Y. Bot. Gard. 9:197. 1955.

Annual or perennial, monoecious or dioecious, erect or lax and
sprawling, the scape 5 cm to 2 m tall, the flower stalk equalling
or shorter than the leaves, drooping; emersed leaves sagittate,
the floating leaves ovate, the submersed leaves reduced to linear
phyllodia; petiole long, usually rather smooth with expanded
base; emersed blade sagittate, prominently nerved, deltoid or
deltoid-orbicular with acute apex, 1–10 cm long, 1–16 cm broad,
basal lobes ovate-lanceolate, acute or acuminate, shorter than
length of blade, 1–12 cm long, 1.0–7.5 cm broad; floating blades
ovate, to 2.5 cm long, the submersed leaves reduced to phyllodia; scape with 3–12 whorls of flowers; peduncle inflated,
smooth, with few flowers; flowers in whorls of 3, the upper
either staminate or pistillate, the lower usually pistillate; bracts
3, fused basally to form a sheath around stem, not prominently
veined, small and hyaline, ovate, acute, to 1 cm long, about
0.5 cm broad; pedicels long, inflated, terete, to 6 cm long, the
lowermost usually becoming reflexed; sepals free, short, persistent, erect and closed, appressed to fruiting head, ovate to ovate-
elliptic with obtuse apex, to 1.3 cm long, to 1.7 cm broad; petals
free, equalling or slightly shorter than sepals, ovate, obtuse,
0.7–15 cm long, 0.5–1.5 cm broad; stamens numerous, to 3 mm
long, the filaments narrowly winged at margin, pubescent, equalling anthers; fruiting head globoid, up to 2 cm in diameter
when mature, very smooth in appearance due to horizontal beak
of achene; achene narrowly cuneate-ovoid, to 3 mm long, to 2.5
mm broad, narrowly winged on both margins, with a well-
developed resin duct at apex of embryo and at the base of horizontal, straight beak; beak about 1 mm long.

5. *Sagittaria calycina* (Arrowleaf). *a*. Leaf, X¼. *b*. Pistillate inflorescence, X⅜. *c*. Staminate inflorescence, X⅜. *d*. Fruiting head, X½. *e*. Achene, X5. *f*. Depauperate plant, X½.

COMMON NAME: Arrowhead.

HABITAT: Marshes, pond margins, shorelines, and shallow water.

RANGE: Ohio to South Dakota, south to New Mexico, Louisiana, and Tennessee; Mexico.

ILLINOIS DISTRIBUTION: Locally throughout central and southern Illinois; apparently absent from the northern counties.

This plant is identified rather easily by the greatly enlarged peduncle and pedicels, the erect sepals which are tightly appressed to the fruiting head, and the leaves which have a wrinkled appearance. As in most of the Sagittarias, the achene is an important characteristic in identification, and should be mature. If immature, the achene can readily be confused with that of S. *latifolia*, but S. *calycina* may be separated by its bracts, peduncle, and pedicel. The pubescent filaments, hyaline bracts, straight long beak of achene, and well-developed resin ducts of the achene readily identify this species.

Both gigantic (to 2 m tall) and dwarfed (5 cm tall) specimens occur.

Although Bogin (1955) chooses to consider this taxon as a subspecies of the primarily South American S. *montevidensis*, I agree with Beal that these two entities should be treated as specifically distinct.

2. Sagittaria rigida Pursh, Fl. Am. Sept. 397. 1814. *Fig. 6.*

Sagittaria heterophylla Pursh, Fl. Am. Sept. 396. 1814, non Schreb. (1811).

Sagittaria heterophylla var. *rigida* (Pursh) Englem. in Gray, Man. Bot. 439. 1856.

Plants monoecious or dioecious, erect, 1.5–8.0 dm tall, the flower stalk equalling or slightly shorter than the leaves, rather slender; petiole long, weak, slightly ridged, with expanded base; blades ovate-elliptic, lanceolate, or hastate, prominently nerved, acute at apex, 2–16 cm long, 3–9 cm broad, or blade lanceolate with acute apex, 8–18 cm long, 1–4 cm broad, or blade hastate with acute apex, 5–15 cm long, 3–9 cm broad, the lobes either 1 or 2, if 2, poorly developed, unequal in length and width, 3–8 cm long, 0.5–1.5 cm broad; peduncle weak, slightly ridged, few-flowered; flowers in whorls of 2–8, the upper

6. *Sagittaria rigida* (Arrowleaf). *a*. Habit (shaded), X$\frac{1}{16}$. *b*. Leaves,
X$\frac{1}{4}$. *c*. Inflorescence, X$\frac{1}{4}$. *d*. Achene, X1$\frac{3}{4}$.

either staminate or pistillate, the lower pistillate; bracts opposite the pedicels, fused about three-fourths entire length,
weak, prominently nerved with a hyaline margin, rather inconspicuous, ovate, obtuse, 2–6 mm long, including the lobes up
to 2 mm long, to 5 mm broad; pedicels of staminate flower long,
slender, terete; pistillate flowers sessile; sepals free, short, with
hyaline margin, persistent and reflexed after flowering, ovate-
lanceolate with rounded or obtuse apex, 0.4–1.0 cm long, to
0.5 cm broad; stamens numerous, to 4 mm long, filaments inflated, densely pubescent or scaly, exceeding length of broad
anther; fruiting head globoid, 2–3 cm in diameter when mature,
prickly in appearance due to erect curved beak of achene;
achene oblongoid, 2.5–7.0 mm long, 2.5–4.0 mm broad, with
a well-developed crested dorsal wing, a small narrow ventral
wing, usually one small facial keel, with resin ducts well-developed at apex of embryo; beak of achene usually terminal,
erect, recurved, 1.0–1.5 mm long; 2n = 22 (Oleson, 1941).

COMMON NAME: Arrowleaf.
HABITAT: Shallow water, muddy sand, in swamps,
margins of ponds, or along waterways.
RANGE: Quebec to Minnesota, south to Nebraska,
Missouri, Tennessee, and Virginia.
ILLINOIS DISTRIBUTION: Relatively uncommon in the
state; it has been found in only 18 counties.
Sagittaria rigida is recognized readily by the sessile
fruiting heads, inconspicuous bracts, and long recurved beak of the achene. The three basic types of
leaves—ovate-elliptic, lanceolate, and hastate—may all be borne
on the same plant. Frequently, however, only one of the leaf
forms may be present. Usually, if the plant is submerged for
any length of time, the unlobed forms develop; if emersed, the
lobed leaves develop.

Several authors state that phyllodia are present in this
species. Of the specimens examined, no trace of phyllodia was
noted in the Illinois specimens.

Bogin (1955) reports that S. *rigida* apparently hybridizes
with S. *graminea*.

3. **Sagittaria graminea** Michx. Fl. Bor. Am. 2:190. 1803. var.
graminea *Fig. 7*.

Sagittaria acutifolia Pursh, Fl. Am. Sept. 2:397. 1814.

Plants monoecious or dioecious, erect, 0.5–5.0 dm tall, the flower stalk equalling or slightly shorter than leaves, rather slender in appearance; petiole long, slender, minutely ridged, with expanded base; blade either entire, hastate, or represented by bladeless phyllodia, prominently nerved, the unlobed blade linear-lanceolate, tapering to acute apex, 2–17 cm long, 3–5 cm broad, the hastate blade with lobes usually unequal in both length and width, up to 2.5 cm long, 2 mm broad, the phyllodia linear-lanceolate, tapering to acuminate tip, to 19 cm long, to 2 cm broad; peduncle slender, minutely ridged, sparsely flowered, the flowers in whorls of 3, upper flowers either staminate or pistillate, lower flowers pistillate; bracts 3, opposite the pedicel, basally connate to fused one-half their entire length, weak, prominently nerved, hyaline in appearance, inconspicuous, ovate, obtuse, 2 mm long, 1–2 mm broad; pedicels long, slender, terete, with staminate and pistillate being same length, 1–3 cm long; sepals basally connate, short, with a hyaline margin, persistent, reflexed in fruit, ovate-lanceolate with obtuse apex, 3–6 mm long, 1–3 mm broad; petals free, greatly exceeding sepals, ovate, narrowing to slender claw at base, obtuse, 1–2 cm long, 1–2 cm broad; stamens numerous, to 3 mm long, the filaments inflated, scaly, shorter than anther; fruiting head globoid, small, 4–8 mm in diameter when mature, smooth in appearance; achene narrowly ovoid, 2 mm long, 1 mm broad, with both ventral and dorsal wings well-developed, 1–3 facial wings with one well-developed, with a well-developed resin duct along one side of embryo, the beak usually oblique to horizontal and borne below summit of achene, short, less than 0.3 mm long; 2n = 22 (Brown, 1947).

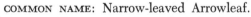

COMMON NAME: Narrow-leaved Arrowleaf.

HABITAT: Swamps, mud, sand, or in shallow water.

RANGE: Labrador to South Dakota, south to Texas and Florida; Cuba.

ILLINOIS DISTRIBUTION: Relatively uncommon in Illinois, where it is found mainly in the central and northern counties.

Sagittaria graminea is most often confused with *S. rigida*, but can easily be separated by the fruiting heads with long pedicels in *S. graminea*. The very

7. Sagittaria graminea var. *graminea* (Narrow-leaved Arrowleaf).   *a.* Habit (left), X⅛.   *b.* Habit (right), X¼.   *c.* Leaf, X¼.   *d.* Pistillate flower, X½. *e.* Achene, X6¼.

short horizontal beak of the achene with 1–3 facial wings, the small, fused, hyaline bracts, and presence of phyllodia all help to separate *S. graminea* from *S. rigida* and the other members of this genus in Illinois.

Two additional varieties of *S. graminea* approach Illinois, but have not as yet been found in Illinois. Variety *platyphylla*, with recurved pedicels of the pistillate heads and an achene beak longer than 3 mm, is a southwestern United States taxon which reaches north to the "boot-heel" of Missouri. Variety *cristata*, with the filaments as long as the anthers, is a taxon of the Great Lakes region, which apparently extends into Iowa and Minnesota.

4. **Sagittaria cuneata** Sheld. Bull. Torrey Club 20:283. 1893.
Fig. 8.

Sagittaria arifolia Nutt. ex J. G. Sm. Ann. Rep. Missouri Bot. Gard. 6:32. 1894.

Plants monoecious or dioecious, erect, to 7 dm tall, the flower stalk usually exceeding the leaves, slender; petiole long, slender, generally smooth, with a slightly expanded base; blade sagittate, prominently nerved, broadly rounded or slightly tapering to acute, bristle-tipped apex, with median constrictions along each margin opposite the attachment of petiole, 6–10 cm long, 1.0–4.5 cm broad, basal lobes broadly lanceolate, tapering to acuminate apex, equalling or slightly exceeding the blade in length; peduncle slender, smooth; inflorescence sparsely flowered, the flowers in whorls of 3, unbranched, the upper predominantly staminate, the lower predominantly pistillate, occasionally both staminate and pistillate at the same node; bracts 3, opposite the pedicel, basally connate, firm, prominently nerved with a subhyaline margin, usually shorter than immature flower head, erect, narrowly lanceolate, acute to acuminate at apex, to 2 cm long, to 5 mm wide; pedicels long, slender, terete, with the staminate slightly longer than pistillate, the staminate up to 2 cm long, the pistillate up to 1 cm long; sepals free, short, persistent and usually reflexed after flowering, ovate with obtuse apex, 8 mm long, 5 mm broad; petals free, equalling or exceeding the sepals, ovate and tapering to a slender claw, obtuse at apex, 8–10 mm long, 5–8 mm broad; stamens numerous, to 5 mm long, the filaments slender, glabrous, 1 mm longer than the anther; fruiting head globoid, to 1.5 cm in

8. *Sagittaria cuneata* (Arrowleaf). *a.* Leaves and inflorescence, X¼. *b.* Fruiting heads, X¼. *c.* Achene, X3¾.

diameter when mature; achene ovoid, up to 2.8 mm long, to 2 mm broad, with small ventral and lateral wings, with a small facial wing, with resin ducts undeveloped, the erect, terminal beak up to 0.4 mm long; 2n = 22 (Brown, 1947, as *S. arifolia*).

COMMON NAME: Arrowleaf.

HABITAT: Mud or water in sloughs, and along waterways.

RANGE: Nova Scotia and Quebec to Northwest Territory and British Columbia, south to California, east to Texas, Illinois, and New York.

ILLINOIS DISTRIBUTION: In central and northern counties; known from 9 counties.

Sagittaria cuneata can easily be distinguished, when mature, by the slender appearance of the plant and the very short, erect, terminal beak of the ovoid achene. The immature specimens are very difficult to separate from *S. brevirostra*.

5. **Sagittaria brevirostra** Mack. & Bush, Missouri Bot. Gard. Rep. 16:102. 1905. *Fig. 9.*

Sagittaria engelmanniana J. G. Sm. ssp. *brevirostra* (Mack. & Bush) Bogin, Mem. N. Y. Bot. Gard. 9(2):244. 1955.

Plants monoecious or dioecious, erect, 4–7 dm tall, the flower stalk equalling or slightly exceeding leaves, robust in appearance; petiole long, stout, ridged, with expanded bases; blade sagittate, hastate, or very rarely entire, prominently nerved, the blade usually tapering to the acute apex, with median constrictions along each margin opposite the attachment of the petiole, 10–22 cm long, 3–12 cm broad, basal lobes lanceolate to ovate-lanceolate, tapering to the acute, bristle-tipped apex, equalling or slightly exceeding the blade in length, 9.5–22.0 cm long, 2.5–7.5 cm broad; peduncle stout, ridged, densely flowered, the flowers in whorls of 3 or 4, the lower whorls occasionally replaced by branches, the upper flowers either staminate or pistillate, the lower flowers entirely pistillate; bracts as many as and opposite the pedicels, free or slightly connate basally, firm, prominently nerved with a hyaline margin, equalling or exceeding the immature flower head, arched-ascending with reflexed tips, lanceolate to ovate-lanceolate, long-attenuate or acute, 1.5–4.0 cm long, 0.5–1.0 cm broad; pedicels long,

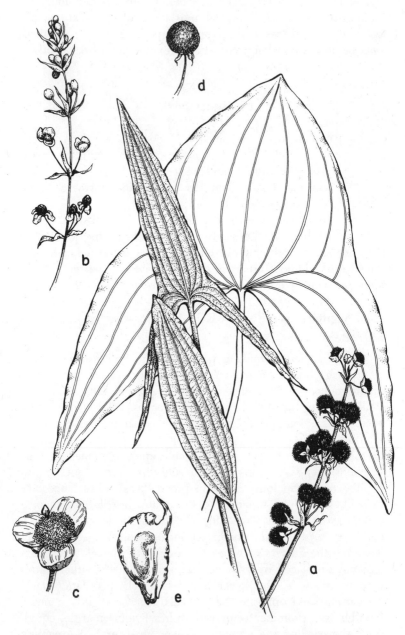

9. *Sagittaria brevirostra* (Arrowleaf). *a.* Leaves and inflorescence, X¼. *b.* Inflorescence, X¼. *c.* Staminate flower, X⅜. *d.* Fruiting head, X½. *e.* Achene, X5.

slender, terete, those of the staminate flowers being slightly longer than those of the pistillate, staminate 1.5–3.0 cm long, pistillate 1–2 cm long; sepals free, short, with hyaline margin, persistent and usually reflexed after flowering, ovate-lanceolate with acute or obtuse apex, 0.6–1.5 cm long, 0.6–1.5 cm broad; petals free, exceeding the sepals, orbicular with a short claw, obtuse, 1–2 cm long, 1.0–1.5 cm broad; stamens numerous, to 5 mm long, filaments slender, glabrous, equalling the anthers; fruiting head globoid, 2–3 cm in diameter when mature; achene cuneate-ovoid, 2.5–3.0 mm long, 1–2 mm broad, with a ventral narrow wing and a broad, crested dorsal wing, with 1–3 facial wings with one usually well developed, with resin ducts not developed at apex of embryo, the broad-based beak usually oblique, rarely erect, recurved at apex, 0.5–1.5 mm long.

COMMON NAME: Arrowleaf.

HABITAT: Sloughs, shorelines, and shallow water.

RANGE: Michigan to South Dakota, south to Texas, Mississippi, and Georgia.

ILLINOIS DISTRIBUTION: Rather common throughout the state.

Sagittaria brevirostra is most often confused with S. *latifolia*, but can easily be distinguished by the long attenuate bracts, the presence of facial wings on the achene, and the beak of the achene with recurved apex. Much confusion can occur when examining immature specimens of S. *brevirostra,* S. *cuneata,* and S. *latifolia.* Even though achenes are not fully developed, the relative position and length of the beak and size and shape of the bracts separate the species adequately.

6. **Sagittaria longirostra** (Micheli) J. G. Sm. Mem. Torrey
   Club 5:26. 1894. *Fig. 10.*

Sagittaria sagittifolia var. *longirostra* Micheli in DC. Monogr. Phan. 3:69. 1881.
Sagittaria longirostra var. *australis* J. G. Sm. in Mohr, Bull. Torrey Club 24:20. 1897.
Sagittaria australis (J. G. Sm.) Small, Fl. SE. U. S. 46. 1903.
Sagittaria engelmanniana ssp. *longirostra* (Micheli) Bogin, Mem. N. Y. Bot. Gard. 9:223. 1955.

Plants monoecious or dioecious, erect, to 6 cm tall, the flower stalk usually a little longer than the leaves; petiole long, stout,

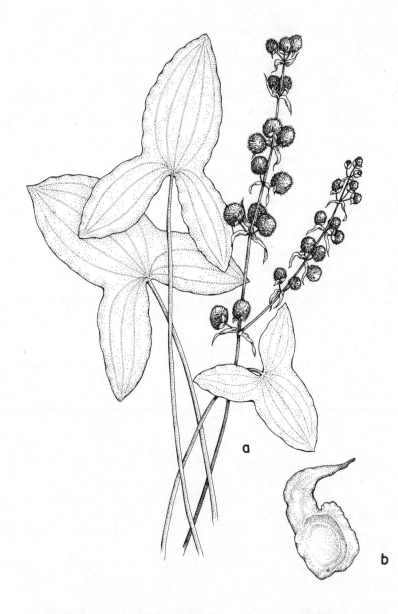

10. *Sagittaria longirostra* (Arrowleaf). *a*. Leaves and inflorescence, X⅛.
b. Achene, X5.

ridged; blade sagittate, prominently nerved, obtuse at apex, to 13 cm long; to 10 cm broad, the basal lobes broad; peduncle stout, ridged, densely flowered, the flowers in whorls of 5–12, the upper flowers staminate or pistillate, the lower flowers usually all pistillate; bracts opposite the pedicels, free or basally connate, firm, arched-ascending, broadly lanceolate, acute at apex, to 2.5 cm long; pedicels rather long, slender, terete, to 2.5 cm long; sepals free, short, persistent and usually reflexed after flowering, to 1.5 cm long, nearly as broad; petals free, a little longer than the sepals, with a short claw; stamens numerous, to 5 mm long, the filaments slender, glabrous; fruiting head about 1.5 cm thick, echinate; achene cuneate-obovoid, 2.3–3.2 mm long, 1.6–2.3 mm broad, with a ventral narrow wing and a broad dorsal wing, with 1–2 facial wings, with resin ducts not developed at apex of embryo, the beak oblique, strongly recurved, 1.0–1.5 mm long.

COMMON NAME: Arrowleaf.

HABITAT: Low, springy woodlands.

RANGE: New York to Illinois, south to Louisiana and Georgia.

ILLINOIS DISTRIBUTION: Known from several springy woodlands in Pope County. First collected in flower, September, 1967, from Long Spring, by Mr. John Schwegman.

Bogin (1955) considers this species to be a subspecies of *S. engelmanniana* from which it differs by its longer bracts and by its more than four flowers in a whorl. I believe that the differences merit specific recognition of this taxon.

This Illinois station marks the northwestern limit in the range of this species.

7. **Sagittaria latifolia** Willd. Sp. Pl. 4:409. 1806. var. **latifolia**
Fig. 11.

Sagittaria simplex Pursh, Fl. Am. Sept. 2:397. 1814.

Sagittaria variabilis Engelm. in Gray, Man. Bot. 461. 1848.

Sagittaria variabilis var. *diversifolia* Engelm. in Gray, Man. Bot. 439. 1856.

Sagittaria latifolia f. *diversifolia* (Engelm.) B. L. Robins. Rhodora 10:31. 1908.

Plants monoecious or dioecious, erect, 4.0–8.3 cm tall, the flower stalk equalling or slightly exceeding the leaves, robust in ap-

11. *Sagittaria latifolia* (Common Arrowleaf). *a.* Leaf, X¼. *b.* Leaves of narrow-leaved form, X¼. *c.* Inflorescence, X¼. *d.* Staminate flower, X½. *e.* Achene, X3.

pearance; petiole long, stout, ridged, with expanded base; blade hastate or sagittate, prominently nerved, usually broadly rounded to the acute apex, with median constrictions along each margin opposite the attachment of the petiole, 6–20 cm long, 5–19 cm broad; basal lobes broadly lanceolate, tapering to the acute or obtuse, bristle-tipped apex, usually equalling or rarely exceeding the blade in length, 5–23 cm long, 2–11 cm broad; peduncle stout, smooth, sparsely flowered, the flowers in whorls of 3, rarely branched at first node, the entire stalk staminate, pistillate, or both; bracts 3, opposite the pedicels, free or slightly connate basally, thin, prominently nerved with hyaline margin, much shorter than the flower head, erect, ovate-lanceolate, obtuse, 1 cm or less long, 1 cm or less broad; pedicels long, slender, terete, with those of the staminate and pistillate flowers about equal in length, to 5 cm long; sepals free, short, with hyaline margin, persistent and reflexed after flowering, ovate, the apex obtuse or shallowly bilobed, 0.5–1.5 cm long, 0.5–1.0 cm broad; petals free, exceeding the sepals, orbicular with short claw, obtuse, 1–2 cm long, 1–2 cm broad; stamens numerous, to 5 mm long; filaments slightly inflated, glabrous, longer than the anthers; fruiting head globoid, 2–3 cm in diameter when mature; achene obovoid, 2.5–3.0 mm long, 3–4 mm broad, with a narrow ventral wing and a broad dorsal wing forming a low crest at apex of achene, without facial wings, with usually one well developed resin duct on facial surface over center of embryo, with the beak of the achene horizontal and straight, to 2 mm long.

COMMON NAME: Common Arrowleaf.
HABITAT: Swamps, sloughs, ponds, shorelines, shallow water, or mud.
RANGE: Nova Scotia to Saskatchewan, south to Colorado, Texas, and Florida; British Columbia, Washington, Oregon, Idaho, California, Arizona; Mexico, through Central America, to Ecuador; Hawaii.
ILLINOIS DISTRIBUTION: More common in the northern and central counties of the state.

In Illinois, S. latifolia is not as common as reported by earlier authors, primarily due to the confusion with S. brevirostra. (See discussion of S. brevirostra.)

The leaf outline of S. latifolia is extremely variable and, therefore, not a good identification character. The illustrations show the extremes in leaf variation. The thin, short, obtuse bracts, mature achene with no facial wings, and the horizontal, straight beak easily distinguish this species.

3. Alisma L. – Water Plantain

Flowers perfect; receptacle flat; sepals 3, green; petals 3, white; stamens 6 or 9; ovaries 10–25, distinct, arranged in a single series; achenes flattened, ribbed only on the back.

This genus is distinguished readily from *Sagittaria* and *Echinodorus* by its ovaries arranged in a single series on the flat receptacle.

A monograph of the genus, prepared by Hendricks (1957), does not seem to depict adequately our species.

KEY TO THE TAXA OF Alisma IN ILLINOIS

1. Petals 3.5–6.0 mm long; flowers at least 7 mm broad; styles at anthesis as long as the ovaries; achenes at least 2.5 mm long_____
   _____1. A. plantago-aquatica var. americanum
1. Petals 1.0–2.5 mm long; flowers at most 3.5 mm broad; styles at anthesis less than one-half as long as the ovaries; achenes less than 2.5 mm long_____2. A. subcordatum

1. Alisma plantago-aquatica L. var. americanum Roem. & Schultes, Syst. 7:1598. 1830. Fig. 12.

Alisma triviale Pursh, Fl. Am. Sept. 1:252. 1814.
Alisma plantago L. var. *triviale* (Pursh) BSP. Prel. Cat. N. Y. Pl. 58. 1888.
Alisma plantago-aquatica L. var. *triviale* (Pursh) Farwell, Ann. Rep. Comm. Parks & Boulev. Det. Det. 11:44. 1900.

Rooted perennial; leaves basal, the blade elliptic to ovate, membranous, to 15 cm long, to 8 cm broad, subacute at apex, rounded to cordate at base, glabrous, the petioles much longer than the blades; scape (primary stalk of inflorescence) erect, bearing a panicle with the secondary branches borne in whorls of 3–10; bracts of inflorescence ovate or narrowly ovate; flowers at least 7 mm broad; sepals obtuse, 3–4 mm long; petals white, 3.5–6.0 mm long; stamens 6–9, at least twice as long as the ovaries; styles at anthesis as long as the 10–25 ovaries; fruiting heads 5–7 mm in diameter; achenes broadly rounded on back, 2.5–3.0 mm long; 2n = 14 (Brown, 1947; Gadella & Kliphuis, 1963; Sokolovskaya, 1963).

12. *Alisma plantago-aquatica* var. *americanum* (Water Plantain).   *a.* Habit,
X¼.   *b.* Achene, X1¾.

COMMON NAME: Water Plantain.

HABITAT: Shallow water, marshes, ditches.

RANGE: Nova Scotia to Quebec and Oregon, south to California, New Mexico, Illinois, West Virginia, and Maryland.

ILLINOIS DISTRIBUTION: Not common; known from 12 counties in the northern half of the state; also Wabash County.

Typical var. *plantago-aquatica,* from Europe, has mostly lilac petals, longer anthers, and longer styles.

For a discussion of *Alisma plantago-aquatica,* see Fernald (1946).

This species is distinguished readily from *A. subcordatum* by its larger flower parts, elongated stamens, and larger achenes.

Pogan (1963) reports 2n = 28 for *Alisma triviale,* here considered synonymous with *A. plantago-aquatica* var. *americanum.*

2. **Alisma subcordatum** Raf. Med. Rep. N. Y. II. 5:362. 1808.

Fig. 13.

Alisma plantago Bigel, Fl. Bost. 87. 1814.

Alisma parviflorum Pursh, N. Am. Sept. 1:252. 1814.

Alisma plantago var. *parviflorum* (Pursh) Torr. Fl. N. & Mid. U. S. 1:382. 1824.

Alisma plantago-aquatica L. var. *parviflorum* (Pursh) Farwell, Ann. Rep. Comm. Parks & Boulev. Det. 11:44. 1900.

Rooted perennial; leaves basal, the blade elliptic to ovate, membranous, to 12 cm long, to 6.5 cm broad, subacute at apex, rounded to cordate at base, glabrous, the petioles much longer than the blades; scape erect, bearing a panicle with much-branched secondary branches borne in whorls of 3–10; bracts of inflorescence lanceolate; flowers at most 3.5 mm broad; sepals obtuse, 1–3 mm long; petals white, 1.0–2.5 mm long; stamens 6–9, slightly longer than the ovaries; styles at anthesis less than one-half as long as the 10–25 ovaries; fruiting heads 3–4 mm in diameter; achenes 1.5–2.5 mm long; 2n = 14 (Pogan, 1963).

13. Alisma subcordatum (Small-flowered Water Plantain). X¼.

COMMON NAME: Small-flowered Water Plantain.
HABITAT: Shallow water, marshes, ditches.
RANGE: Vermont to Minnesota, south to Florida and Texas; Mexico.
ILLINOIS DISTRIBUTION: Common; probably in every county, but not collected as yet in the south-central counties.

The smaller flowers and achenes distinguish this species from A. *plantago-aquatica* var. *americanum*.

The flowers appear from mid-July until the end of August.

HYDROCHARITACEÆ – FROG'S-BIT FAMILY

Flowers perfect or unisexual; perianth parts 3 or 6, in 1 or 2 usually distinguishable series; stamens 1–12; ovary 1, inferior (sometimes interpreted as superior in *Elodea*).

St. John (1965) proposes to resurrect the family Elodeaceae, including *Elodea*, on the basis of the interpretation of a superior ovary in this genus.

All members of this family in Illinois live in or around water.

KEY TO THE GENERA OF Hydrocharitaceae **IN ILLINOIS**

1. Leaves borne along the stem, the longest not more than 3 cm, the broadest not more than 5 mm_____1. *Elodea*
1. Leaves all basal, 2–200 cm long, 5–25 mm broad.
    2. Leaves narrow, elongate; fertile stamens 2, the filaments free___
    _____2. *Vallisneria*
    2. Leaves ovate to orbicular; fertile stamens 6–12, the filaments united into a column_____3. *Limnobium*

1. *Elodea* MICHX. – Waterweed

Flowers perfect or unisexual, usually dioecious, produced from a bilobed spathe; staminate spathes 1- to 4-flowered, the flowers with 3 sepals, 3 petals, 9 (–10) stamens with the filaments free or united into a column; perfect flowers with 3 stamens; pistillate flowers with 3 sepals united below into a tube, 3 petals, 1 ovary with 1 locule; fruit few-seeded.

The perianth tube in the pistillate and perfect flowers becomes greatly elongated (to 30 cm) at anthesis so that the lobes of the perianth reach the surface of the water.

The generic name, *Elodea,* is conserved over the earlier name, *Anacharis.* St. John (1965) reestablished the genus *Egeria* for our first species below, distinguishing it from *Elodea* principally by its 2- to 4-flowered staminate spike and its free stamens. This separation appears to be unnecessary.

KEY TO THE SPECIES OF Elodea IN ILLINOIS

1. Leaves in whorls of 4 or 6, generally more than 2 cm long; petals 9–12 mm long_____1. *E. densa*
1. Leaves opposite or in whorls of 3, generally less than 2 cm long; petals minute or up to 5 mm long.
   2. Staminate flowers pedicellate, with sepals 3.5–5.0 mm long; leaves 1–5 mm broad_____2. *E. canadensis*
   2. Staminate flowers sessile, liberated at maturity, with sepals 1.5–2.0 mm long; leaves smaller, usually 0.5–1.5 mm broad___ _____3. *E. nuttallii*

1. **Elodea densa** (Planch.) Caspary, Monatsber. Kgl. Preuss. Akad. Wissensch. 1857:48. 1857. *Fig. 14.*

Egeria densa Planch. Ann. Sci. Nat. Bot. III, 11:80. 1849.
Anacharis densa (Planch.) Vict. Contr. Lab. Bot. Univ. Montreal 18:41. 1931.
Pistillate plants not observed.
Staminate plants: leaves linear-lanceolate, subacute, mostly in whorls of (4–) 6, over 2 cm long, 2–5 mm broad; sepals 5–8 mm long; petals 9–12 mm long, white; stamens apparently 9; n = 24 (Matsuura & Sutô, 1935).

COMMON NAME: Elodea.
HABITAT: Quiet water.
RANGE: Native to South America; occasionally introduced in the United States.
ILLINOIS DISTRIBUTION: Very rare; known only from Williamson County, where it is abundant in Madison Pond (*H. L. Hester s.n.*) and from Franklin County, West Frankfort Lake (*D. L. Tindall s.n.*).
This species is adventive in Illinois. It is distinguished readily by its large flowers and its leaves in whorls of 4–6. The method of introduction into Illinois waters is at this time unknown, although it is suspected to be the result of aquarium disposals.

14. Elodea densa (Elodea).   *a.* Habit, X¼.   *b.* Flower (staminate), X1¼.

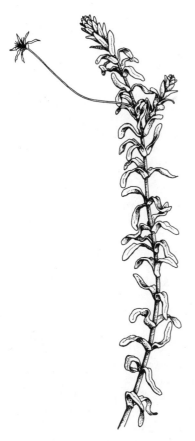

15. Elodea canadensis (Elodea). X½.

2. Elodea canadensis Rich. in Michx. Fl. Bor. Am. 1:20. 1803. *Fig. 15.*

Udora canadensis (Rich.) Nutt. Gen. N. Am. Pl. 2:242. 1818.
Anacharis canadensis Planch. Ann. & Mag. Nat. Hist. II, 1:86. 1848, not based on *E. canadensis* Rich. in Michx.

Pistillate plants: leaves broadly lanceolate, obtuse or subacute, the middle and upper in whorls of 3, the lower opposite, to 13 mm long, to 4 (–5) mm broad; perianth tube thread-like, to 15 cm long; sepals about 2 mm long; petals 2.3–2.6 mm long, white; staminodia 3, very slender, less than 1 mm long; stigmas 3, bilobed; ovary 1, broadly lanceoloid, 2.5–3.0 mm long; fruit 5–6 mm long, ovoid.

Staminate plants: leaves linear or linear-lanceolate, acute, the middle and upper in whorls of 3, the lower opposite, to 10 mm long, to 3.5 mm broad; perianth tube thread-like; sepals 3.5–5.0 mm long; petals 4–5 mm long, white; stamens 9.

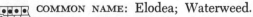

COMMON NAME: Elodea; Waterweed.

HABITAT: Quiet water.

RANGE: Quebec to British Columbia, south to California, Oklahoma, and Alabama.

ILLINOIS DISTRIBUTION: Occasional in the northern half of the state; rare in the southern half.

The staminate plants are rarely seen and collected. This species sometimes is confused with *E. nuttallii,* a species generally with more slender leaves and smaller flowers.

3. **Elodea nuttallii** (Planch.) St. John, Rhodora 22:27. 1920. *Fig. 16.*

Serpicula occidentalis Pursh, Fl. Am. Sept. 1:33. 1814, nomen illeg.

Anacharis nuttallii Planch. Ann. & Mag. Nat. Hist. II. 1:86. 1848.

Elodea occidentalis (Pursh) St. John, Rhodora 22:27. 1920, based on illeg. basionym.

Anacharis occidentalis (Pursh) Vict. Contr. Lab. Bot. Univ. Montreal 18:40. 1931, based on illeg. basionym.

Pistillate plants: leaves linear to linear-lanceolate, more or less acute, the middle and upper leaves in whorls of 3 (rarely 4), the lower opposite, to 12 mm long, to 1.5 mm broad; sepals about 1 mm long; petals 1.2–1.5 mm long, whitish; staminodia minute, about 0.5 mm long; stigmas 3, bilobed; ovary 1, lanceoloid, 1.5–2.5 mm long; fruit 2–4 mm long, lanceoloid.

Staminate plants: leaves as in the pistillate plants; flowers sessile, liberated at maturity and floating free on the surface of the water; sepals about 2 mm long, reddish; petals minute, at most 0.5 mm long; stamens 9.

COMMON NAME: Elodea; Waterweed.

HABITAT: Quiet water.

RANGE: Maine to Missouri and North Carolina; Idaho.

ILLINOIS DISTRIBUTION: Occasional in the northern counties; not common in the southern counties.

The unique feature of this species is the staminate flower which is liberated at maturity and which floats to the surface of the water. The blades, particularly in the pistillate plants, seem to be narrower than those of *E. canadensis.*

16. Elodea nuttallii (Elodea). X½.

2. *Vallisneria* L. – Eelgrass

Flowers unisexual, dioecious, produced from a spathe; sepals 3, unequal in size in the staminate flowers, equal in the pistillate flowers but united to form a tube; fertile stamens 2; ovary 1; fruit many-seeded; leaves very long, ribbon-like.

Reproduction in this genus is curious. The male flowers, while in bud, are liberated and float to the surface of the water where the pistillate flowers have come due to rapid elongation of the pedicel. Following fertilization, the pistillate pedicel coils, pulling the fruit beneath the water where it matures.

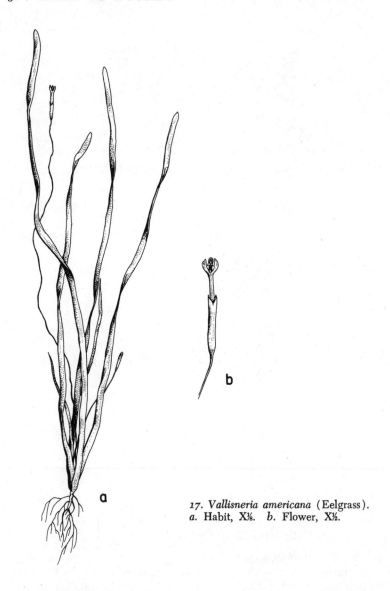

17. *Vallisneria americana* (Eelgrass).
a. Habit, X⅛. *b*. Flower, X½.

1. Vallisneria americana Michx. Fl. Bor. Am. 2:220. 1803.
Fig. 17.

Leaves membranous, to 2 m long, 5–20 mm broad, obtuse; pistillate flowers white, 2–3 cm long; staminate flowers smaller; fruits cylindrical, 5–12 cm long.

COMMON NAME: Eelgrass.

HABITAT: Quiet water.

RANGE: Nova Scotia to Manitoba, south to Texas and Florida.

ILLINOIS DISTRIBUTION: Occasional; restricted to the northern one-half of the state.

Vallisneria americana may occur at considerable depths beneath the surface of the water.

3. Limnobium RICH. – Frog's-bit

Flowers unisexual, dioecious, produced from a spathe; sepals 3, united; petals 3, free or nearly so; filaments united into a column, with 6–12 anthers; ovary 1, inferior, with 6–9 locules; fruit fleshy, many-seeded; leaves broad.

Only the following species occurs in Illinois.

1. Limnobium spongia (Bosc) Steud. Nom. Bot. 1. 1841.

Fig. 18.

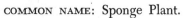

Hydrocharis spongia Bosc, Ann. Mus. Paris 9:396. 1807.

Usually rooted perennial; leaves floating, to 7 cm long, nearly as wide, ovate, subacute at apex, rounded to cordate at base, glabrous, green above, usually purplish below, the blade, at least when young, bearing spongy cells; spathes 1-leaved in the staminate flowers, 2-leaved in the pistillate flowers; peduncle of staminate flower 4–9 cm long, of pistillate flower 2.5–3.5 cm long; sepals 3, oblong, 4–7 mm long; petals 3, linear, 4–7 mm long; fruit globoid, fleshy or soon becoming dry, bluish, about 5–6 mm in diameter.

COMMON NAME: Sponge Plant.

HABITAT: Swamps.

RANGE: Along the coast from New York to Florida, thence to Texas; Illinois; Missouri; Kentucky.

ILLINOIS DISTRIBUTION: Rare; known only from Alexander, Johnson, Massac, and Union counties.

The unique, spongy-inflated cells of the young leaf blades keep the plant buoyed up and floating on the surface of the water. In deeper waters, the plants may be free-floating, rather than rooted. The flowers, which appear in late July and throughout August, are white.

George Vasey collected this plant from the LaRue Swamp of Union County as early as 1862.

18. *Limnobium spongia* (Sponge Plant).   *a.* Habit, X¼.   *b.* Leaf, X⅜.   *c.* Floating leaf, X⅜.   *d.* Staminate flower, X⅜.   *e.* Pistillate flower, X⅜.   *f.* Habit (shaded), X½.

Order Zosterales

Flowers bisexual or unisexual; perianth composed of two scarcely distinguishable series of three members each, or absent; stamens 1, 2, 4, or 6; carpels 3, 4, or 6, superior, united at first, distinct or nearly so at maturity; fruit composed of achenes, drupes, or follicles; leaves terete or flat, basal, alternate, or opposite.

JUNCAGINACEÆ – ARROW-GRASS FAMILY
Flowers perfect; perianth composed of two scarcely distinguishable series of three members each; stamens 6; carpels 3 or 6, united at first, distinct or nearly so at maturity; fruit composed of 3 or 6 follicles; leaves terete.

1. *Triglochin* L. – Arrow-grass

Flowers perfect; sepals 3, greenish, deciduous; petals 3, greenish, deciduous; stamens 6 (in Illinois species); carpels united

61

at first, later separating and developing into 3 or 6 follicles; ovule 1; inflorescence a spike-like raceme, without bracts; leaves all basal.

The contracted racemes, without bracts, and the absence of cauline leaves distinguish this genus from *Scheuchzeria*.

KEY TO THE SPECIES OF Triglochin IN ILLINOIS

1. Carpels 6, the axis between them slender_____1. *T. maritimum*
1. Carpels 3, the axis between them broadly 3-winged_____
   _____2. *T. palustre*

1. **Triglochin maritimum** L. Sp. Pl. 339. 1753. *Fig. 19.*

Perennial from a stocky, non-stoloniferous rhizome; leaves basal, terete, linear, to 50 cm long, 1–3 mm broad, erect, glabrous; scape usually less than 10 dm tall, bearing a bractless, spike-like raceme, naked below; sepals and petals each 3, greenish, deciduous, 1–2 mm long; stamens 6; carpels 6, rounded at base, sharply angled on the margins, the axis between them slender; follicles 6; 2n = 12 (Löve & Löve, 1944); 2n = 24 (Sorsa, 1963).

COMMON NAME: Arrow-grass.

HABITAT: Sandy shores, swamps, or wet ditches.

RANGE: Labrador to Alaska, south to California, New Mexico, Nebraska, Illinois, and New Jersey; Mexico; Europe; Asia; Africa; Patagonia.

ILLINOIS DISTRIBUTION: Not common; known only from the extreme northeastern counties; also Peoria County. The carpels, which are slightly coherent at first but are separate at maturity, are six in number, distinguishing this species from *T. palustre*, which has three carpels.

2. **Triglochin palustre** L. Sp. Pl. 338. 1753. *Fig. 20.*

Perennial from a stocky rootstock, with elongated, bulb-bearing stolons; leaves basal, terete, linear, to 30 cm long, 1–2 mm broad, erect, glabrous; scape usually less than 60 cm tall, bearing a bractless, spike-like raceme, naked below; sepals and petals each 3, greenish, deciduous, 1.0–1.5 mm long; stamens 6; carpels 3, sharp-pointed at base, the axis between them broadly 3-winged; follicles 3; 2n = 24 (Löve & Löve, 1944).

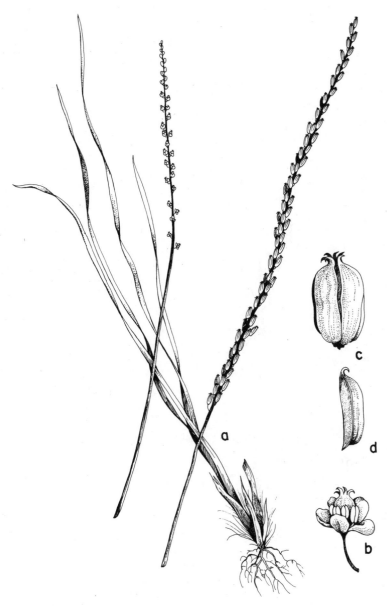

19. Triglochin maritimum (Arrow-grass). *a.* Habit, inflorescence, and fruiting branch, X¼. *b.* Flower, X2½. *c.* Cluster of fruits, X6¼. *d.* Single fruit, X6¼.

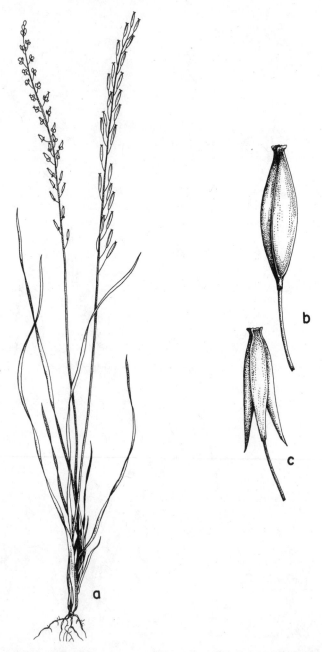

20. *Triglochin palustre* (Arrow-grass). *a.* Habit, X¼. *b.* Flower, X6¼. *c.* Follicles, X6¼.

COMMON NAME: Arrow-grass.

HABITAT: Low situations.

RANGE: Greenland to Alaska, south to California, Nebraska, Illinois, and New York; Europe; Asia; South America.

ILLINOIS DISTRIBUTION: Not common; known only from the northeastern counties; also Peoria and Tazewell counties.

The number of carpels is a reliable character which easily separates this species from *T. maritimum*.

2. Scheuchzeria L. – Arrow-grass

Flowers perfect; sepals 3, greenish, deciduous; petals 3, greenish, persistent; stamens 6; carpels free except at base, developing into 3 follicles; ovules 2; inflorescence loosely racemose, bracteate; leaves basal and alternate.

The distinguishing characters of this genus are the loosely racemose, bracteate racemes and the presence of cauline leaves. Only the following taxon occurs in Illinois.

1. **Scheuchzeria palustris** L. var. **americana** Fern. Rhodora 25:178. 1923. *Fig. 21.*

Scheuchzeria americana (Fern.) G. N. Jones, Fl. Ill. 44. 1945.

Perennial from a creeping, jointed rootstock; leaves terete, linear, to 30 cm long, 1–3 mm broad, open at the tip, glabrous, the basal ones clustered, the cauline ones 1–3, scattered; sheaths at base of leaf enlarged, partly sheathing the stem; raceme bracteate, the lowest bract leaf-like, the upper sheathlike; sepals ovate-lanceolate, acute, 2–3 mm long; petals ovate-lanceolate, acute, 2–3 mm long; fruits 6–10 mm long, with a curved beak; seeds narrowly ellipsoid, 4–5 mm long, black; $2n = 22$ (Manton, 1949).

COMMON NAME: Arrow-grass.

HABITAT: Bogs.

RANGE: Newfoundland to Washington, south to California, Illinois, and New Jersey.

ILLINOIS DISTRIBUTION: Rare; known from two extreme northeastern counties and two west-central counties.

Typical var. *palustris*, which is Eurasian, differs from var. *americana* in its slightly smaller, nearly beakless follicles and its slightly smaller seeds.

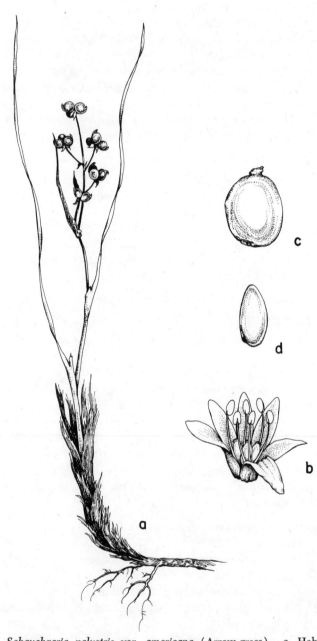

21. *Scheuchzeria palustris* var. *americana* (Arrow-grass). *a.* Habit, X¼.
b. Flower, X3¾. *c.* Follicle, X1¾. *d.* Seed, X1¾.

After examining material of S. *palustris* from its entire range, I have concluded that the above differences noted in the American specimens are not significant enough to justify specific segregation.

POTAMOGETONACEÆ – PONDWEED FAMILY

Only the following genus occurs in Illinois.

1. *Potamogeton* L. – Pondweed

Perennial aquatic herbs; stems simple or branched; leaves 2-ranked, alternate, sometimes di- or tri-morphic, entire or denticulate; stipules sheathing; spikes pedunculate, continuous to moniliform; perianth absent (some botanists consider 4 sepals to be present); stamens 4, with sepaloid connectives; carpels 4, each with a single ovule; fruit drupaceous when fresh, but appearing like a beaked achene after drying.

The species of the genus *Potamogeton* are identified more easily from fresh material than from dried. Characters which should be noted particularly when the specimens are collected are the nature of the stipules and of the submersed leaves. Characters of the fruit may be studied just as well from dried specimens.

Considerable difficulty is encountered in identification of some of the species. This is due in part to the great degree of variation exhibited by some species and in part to the hybridization potential of others.

Several extensive treatments of the genus have been prepared. Most important are those by Graebner (1907) and Hagström (1916), who paid particular attention to the classification of the genus and who are responsible for the numerous subgeneric categories. American works have been by Morong (1893), Taylor (1909), Fernald (1932), and Ogden (1943). Fernald's treatment of the linear-leaved species and Ogden's treatment of the broad-leaved species are excellent and have been relied upon heavily in the preparation of the following account. Ellis' work (1963) on the Illinois species has been used extensively.

KEY TO THE TAXA OF Potamogeton IN ILLINOIS

1. Leaves septate, 0.3–0.5 (–1.0) mm broad; peduncles flexuous. Subgenus Coleogeton_____1. *P. pectinatus*

1. Leaves not septate, some or all over 0.5 mm broad; peduncles rigid. Subgenus Potamogeton.
  2. Leaves uniform.
    3. Stipules adnate to leaf-base; leaves auriculate at base. Section Adnati_____2. *P. robbinsii*
    3. Stipules free from leaf-base; leaves rounded, cordate, or tapering at base, never auriculate. Section Axillares.
      4. Leaves sharply serrulate; fruits 5–6 mm long. Subsection Crispi_____3. *P. crispus*
      4. Leaves entire or minutely denticulate; fruits 1.6–5.0 mm long.
        5. Leaves 5–30 mm broad, rounded or cordate at base.
          6. Leaves entire; fruits 4–5 mm long, keeled. Subsection Praelongi_____4. *P. praelongus*
          6. Leaves minutely denticulate; fruits 2.5–3.5 mm long, not keeled. Subsection Perfoliati_____
          _____5. *P. richardsonii*
        5. Leaves up to 5 mm broad, tapering to base.
          7. Leaves with 15–35 nerves; spikes with 6–11 whorls of flowers; stems flattened; fruit quadrate, 4–5 mm long. Subsection Compressi_____6. *P. zosteriformis*
          7. Leaves with 3–7 nerves; spikes with 1–5 whorls of flowers; stems not flattened; fruit obovoid to ovoid, 1.9–3.0 mm long. Subsection Pusilli.
            8. Stipules connate, forming cylinders.
              9. Spikes subcapitate, 2–5 mm long; fruits keeled; sepaloid connectives 0.5–1.0 mm long; leaves glandless at base_____7. *P. foliosus*
              9. Spikes elongate, more or less interrupted, 6–15 mm long; fruits rounded on back; sepaloid connectives 1.2–2.5 mm long; leaves biglandular at base.
                10. Leaves 5- to 7-nerved, thin, translucent _____8. *P. friesii*
                10. Leaves 3-nerved, firm, opaque.
                  11. Stipules chartaceous, white, becoming fibrous_____9. *P. strictifolius*
                  11. Stipules membranous, olive, not becoming fibrous_____10. *P. pusillus*
            8. Stipules free_____11. *P. berchtoldii*
2. Leaves of two kinds, the floating usually shorter and broader than the submersed.

12. Leaves entire.
13. Submersed leaves up to 2 mm broad, 1- to 5-nerved.
14. Spikes uniform; stipules free from base of leaf;
fruits 1.6–5.0 mm long.
15. Submersed leaves 0.1–0.5 mm broad, 1- to
3-nerved; fruits 1.6–2.2 mm long. Subsection
Javanici_____12. *P. vaseyi*
15. Submersed leaves 0.5–2.0 mm broad, 3- to
5-nerved; fruits 3–5 mm long. Subsection Na-
tantes_____13. *P. natans*
14. Spikes of two or more kinds, those in the axils of
the submersed leaves subgloboid, those in the axils
of the floating leaves elongate; stipules adnate to
base of leaf; fruits 1.0–1.5 mm long. Subsection
Hybridi_____14. *P. diversifolius*
13. Submersed leaves 5–75 mm broad, 7- to 21-nerved.
16. Stems compressed; submersed leaves linear, 5–10
mm broad; fruit capped with a minute, tooth-like
beak. Subsection Nuttalliana_____15. *P. epihydrus*
16. Stems terete; submersed leaves lanceolate to ovate,
10–75 mm broad; fruit capped with a prominent
beak. Subsection Amplifolii.
17. Fruiting spike 4 cm long or longer; stipules of
submersed leaves subpersistent; fruit tapering
to base; some nerves of floating leaves more
prominent than other nerves_____
_____16. *P. amplifolius*
17. Fruiting spike 2.0–3.5 cm long; stipules of submersed leaves falling away early; fruit rounded
at base; all nerves of floating leaves uniform___
_____17. *P. pulcher*
12. Leaves minutely denticulate. Subsection Lucentes (including Subsection Nodosi).
18. Fruits (including beak) 3.5–4.3 mm long_____
_____18. *P. nodosus*
18. Fruits (including beak) 1.7–3.5 mm long.
19. Stems simple or once-branched; stipules strongly
keeled.
20. Floating leaves 2.0–6.5 cm wide; fruits normally developed_____19. *P. illinoensis*
20. Floating leaves 0.5–1.0 cm wide; fruits not ma-
turing_____20. *P.* × *hagstromii*

19. Stems repeatedly branched; stipules faintly keeled.
  21. Floating leaves elliptic to ovate, the petioles usually as long as or longer than the blades; stipules obtuse.
    22. Submersed leaves acute at apex; fruit strongly 3-keeled, well-developed_____ _____21. *P. gramineus*
    22. Submersed leaves cuspidate at apex; fruit obscurely 3-keeled, poorly developed____ _____22. *P.* × *spathulaeformis*
  21. Floating leaves oblong- to linear-lanceolate, the petioles shorter than the blades; stipules acuminate_____23. *P.* × *rectifolius*

1. Potamogeton pectinatus L. Sp. Pl. 127. 1753. *Fig. 22.*

Rhizome slender, fibrous, bearing small tubers. Stem filiform, profusely branched, sometimes attaining a length of nearly one meter. Leaves uniform, submersed, filiform, sharply pointed at the apex, tapering to the somewhat broader sheathing base, 2.0–7.5 cm long, 0.3–1.0 mm broad, the margins entire; stipules connate more than half their length, 2–5 cm long. Spikes interrupted, moniliform, composed of 2–6 whorls of flowers, in fruit 1.5–5.0 cm long; flowers sessile; sepaloid connectives suborbicular to broadly rhombic, 0.7–1.0 mm long, short-clawed. Fruits reniform, smooth and weakly one-keeled on the back, 2.5–4.2 mm long (excluding the beak), 1.0–2.5 mm broad, with a short, curved beak near the ventral margin, brown.

COMMON NAME: Fennel-leaved Pondweed.
HABITAT: Usually calcareous water.
RANGE: Newfoundland to Alaska, south to Texas and Florida; Mexico.
ILLINOIS DISTRIBUTION: Throughout the state; probably in every county.
The septate leaves and flexuous peduncles are sufficient characters to justify the placing of this species into subgenus Coleogeton.

Potamogeton pectinatus may be recognized from all other pondweeds in Illinois by its interrupted spike and its densely branched stems. When *P. pectinatus* is present in a pond, it is usually the most prolific species. It may be possible

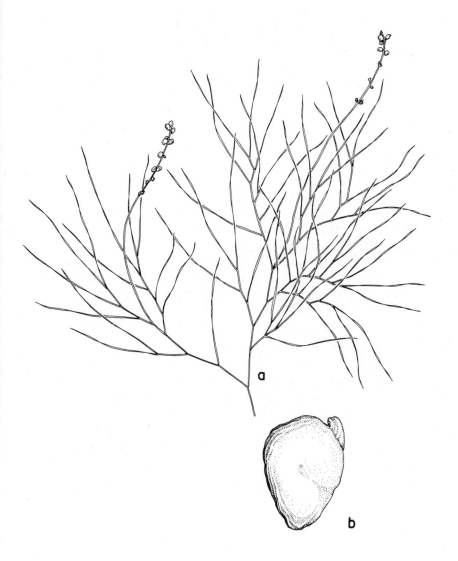

22. *Potamogeton pectinatus* (Fennel-leaved Pondweed). *a.* Habit, X¼.
b. Fruit, X5.

to mistake this species for the lesser branched *P. strictifolius* or *P. friesii*.

Venation in the leaves is highly variable and is not a good character in this species. Leaf shape is constantly filiform.

Reports by Hill (1896), Pepoon (1927), and Jones, *et al.* (1955) of *Potamogeton interruptus* Kitaibel, based on a sterile collection from South Chicago (Cook County) by Hill in 1881, are extremely questionable. This author believes the specimen should be referred to *P. pectinatus*.

2. Potamogeton robbinsii Oakes, Hovey's Mag. 7:180. 1841.

Fig. 23.

Rhizome very slender and long-fibrous, not tuber-producing. Stem simple or profusely branched, sometimes attaining a length of over one meter, rooting at the nodes. Leaves submersed, adnate to the lower half of the stipules, those of the sterile stems crowded, linear to lanceolate, stiff, acute at the apex, auriculate at the clasping base, 2–80 cm long, 1–8 mm broad, 20- to 60-nerved, with a serrulate margin, those of the flowering stems remote and smaller; stipules 1.0–2.5 cm long, sheathing for one-half of their length, whitish, becoming fibrous. Spikes interrupted, cylindrical, branching, to 2 cm long; flowers sessile; sepaloid connectives suborbicular to rhombic, short-clawed. Fruits flattened, obovoid, smooth, with a prominent keel on the back, 4–5 mm long (excluding the beak), 2.7–3.3 mm broad, with a central beak up to 1 mm long; n = 26 (Stern, 1961).

COMMON NAME: Pondweed.

HABITAT: Stagnant water.

RANGE: New Brunswick to British Columbia, south to Oregon, Illinois, and Alabama.

ILLINOIS DISTRIBUTION: Known only from Lake County (Cedar Lake, *Richardson s.n.;* Bang's Lake, Wauconda, *Hill s.n.;* Lake Zurich, *Hill s.n.*).

Potamogeton robbinsii can be immediately recognized by its stipules which are adnate to the leaf base, by its branching, interrupted inflorescence, and by its long, stiff, linear leaves.

The length of the leaves is extremely variable. Fruits are rarely observed.

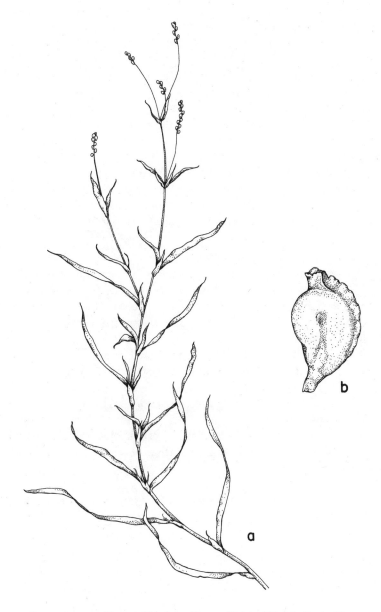

23. *Potamogeton robbinsii* (Pondweed). *a.* Habit, X⅛. *b.* Fruit, X3.

3. **Potamogeton crispus** L. Sp. Pl. 126. 1753. *Fig. 24.*

Rhizome stout, creeping, unspotted. Stem compressed, usually branched, sometimes simple, 0.5–2.5 mm in diameter. Leaves uniform, submersed, broadly linear to oblong, broadly rounded to subacute at the apex, tapering to the sub-clasping base, 1.3–7.5 (–10.0) cm long, 0.2–1.1 cm broad, reddish-green, the margins finely and irregularly dentate; stipules slightly adnate at base, papery, becoming frayed early above. Spikes compact at first, becoming somewhat interrupted, cylindrical, composed of 3–5 whorls of flowers, in fruit 1–2 cm long, 1.0–1.3 cm thick; flowers sessile or short-pedicellate, the pedicels never exceeding 0.5 mm; sepaloid connectives orbicular, (0.6–) 1.2–2.0 mm long, short-clawed, anthers 0.7–1.3 mm long. Fruits ovoid, strongly and obtusely keeled with a small tooth near the base, 2.0–3.5 mm long (excluding the beak), 1.4–2.8 mm wide, with a straight or incurved beak 2–3 mm long, greenish or brownish. Winter buds firm, 1–2 cm thick; n = 26 (Palmgren, 1939).

COMMON NAME: Pondweed.

HABITAT: Muddy to calcareous ponds and streams.

RANGE: Introduced in North America, from southern Quebec and southern Ontario to Minnesota, south to Nova Scotia, Virginia, and Missouri; Oklahoma; California.

ILLINOIS DISTRIBUTION: Restricted to several counties in the northern half of the state, where it is introduced.

The irregularly dentate, crisped margins of the leaves readily distinguish this species. The fruits, which measure about 5–6 mm in length, are the longest in the genus. *Potamogeton crispus* is the only representative of subsection Crispi in Illinois.

4. **Potamogeton praelongus** Wulfen in Roem. Arch. Bot. 3:331. 1805. *Fig. 25.*

Rhizome rather stout, whitish, suffused with red. Stem slender, to 4 mm in diameter, simple or sparingly branched, flexuous, whitish to olive-green. Leaves uniform, all submersed, broadly lanceolate, obtuse and cucullate at the apex, rounded or cordate at the partly clasping base, 2–18 (–36) cm long, 0.8–3.0 cm broad, 13- to 25-nerved, with 3–7 of the nerves more promi-

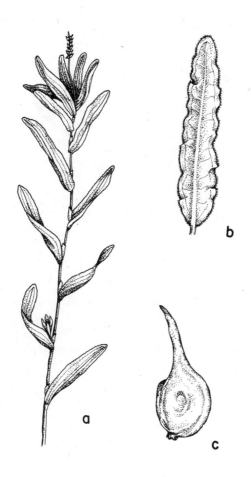

24. *Potamogeton crispus* (Pondweed). *a*. Habit, X⅛. *b*. Leaf, X½. *c*.
Fruit, X3.

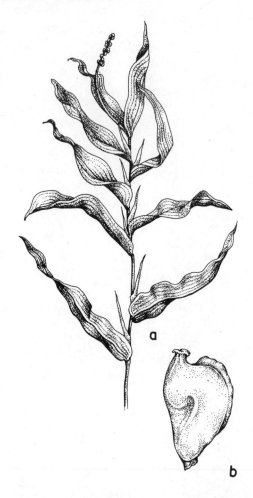

25. Potamogeton praelongus (Pondweed). *a.* Habit, X⅖. *b.* Fruit, X3.

nent, sessile, the margins entire; stipules mostly persistent, more or less oblong, 2.5–6.0 (–10.0) cm long, white, strongly nerved, without keels. Spikes interrupted, sometimes moniliform, cylindrical, with 5–10 (–12) whorls of flowers, in fruit 3–5 cm long, 1.0–1.5 cm broad; flowers sessile; sepaloid connectives suborbicular to rhombic, 1.7–3.0 mm broad, short-clawed. Fruits obovoid, rounded and prominently 1-keeled on the back, 4–5 mm long (excluding the beak), 2.5–4.0 broad, with a thick beak up to 1 mm long; n ÷ 26 (Palmgren, 1939).

COMMON NAME: Pondweed.

HABITAT: Lakes and rivers.

RANGE: Labrador to Alaska, south to California, northern Illinois, and New Jersey.

ILLINOIS DISTRIBUTION: Confined to three counties in extreme northeastern Illinois.

Potamogeton praelongus is characterized by its whitish, flexuous stems, its large fruits, its large, cucullate, ovate leaves, and its conspicuous whitish stipules. It is the only Illinois representative of subsection Praelongi.

5. **Potamogeton richardsonii** (Benn.) Rydberg, Bull. Torrey Club 32:599. 1905. *Fig. 26.*

Potamogeton perfoliatus var. *lanceolatus* Robbins in Gray, Man. Bot. 5:488. 1867, non Blytt (1861).

Potamogeton perfoliatus var. *richardsonii* Benn. Journ. Bot. 27:25. 1889.

Rhizome thick, whitish or yellowish, unspotted. Stem slender, to 2.5 mm in diameter, simple to branched. Leaves all submersed, the lower ovate to ovate-lanceolate, the upper becoming lanceolate, obtuse to subacute at the apex, cordate at the partly clasping base, 1.5–14.0 cm long, 0.5–3.5 cm broad, 7- to 33-nerved, with 3–7 nerves more prominent, with the margins denticulate; stipules ovate to lanceolate, obtuse, 1.0–2.7 cm long, obtuse, whitish, strongly nerved, without keels. Spikes mostly interrupted and moniliform, sometimes continuous, cylindrical, composed of 6–12 whorls of flowers, in fruit 1.5–5.0 cm long, 1 cm broad; flowers sessile; sepaloid connectives suborbicular to rhombic, 1.5–2.5 mm broad, shortclawed. Fruits obovoid, rounded and usually keelless on the back, 2.5–3.5 mm long (excluding the beak), 2.0–3.3 mm broad, with a beak up to 1 mm long, brownish; n = 26 (Stern, 1961).

COMMON NAME: Pondweed.

HABITAT: Lakes and rivers.

RANGE: Labrador to Alaska, south to California, northern Illinois, and New York.

ILLINOIS DISTRIBUTION: Restricted to Lake, Cook, and McHenry counties in extreme northeastern Illinois. The report from Kankakee County (Jones, *et al.*, 1955) could not be verified.

The numerous, coarsely-nerved, clasping leaves and the persistent, whitish, fibrous stipules characterize this species.

26. Potamogeton richardsonii (Pondweed). *a.* Habit, X⅛. *b.* Fruit, X3.

This species has been referred to by earlier botanists as *P. perfoliatus,* but this species occurs far to the north of Illinois. Robbins and Bennett first treated *P. richardsonii* as a variety of *P. perfoliatus.* *Potamogeton perfoliatus* differs by its leaves with less prominent nerves and by its extremely delicate stipules.

6. Potamogeton zosteriformis Fern. Mem. Amer. Acad. 17: 36. 1932. *Fig. 27.*

Rootstock slender, fibrous. Stem usually branched, flattened, to 3 mm broad, constricted at the nodes. Leaves uniform, submersed, linear, elongate, obtuse to acuminate at apex, clasping at base, 6–12 (–20) cm long, 2–5 mm broad, 15- to 35-nerved, with three nerves more prominent, sessile; stipules firm, 1.5–4.0 cm long, the lower obtuse, the upper acuminate, becoming fibrous. Spikes nearly continuous, cylindric, composed of 6–11 whorls of flowers, in fruit 1.5–4.0 cm long; flowers sessile; sepaloid connectives suborbicular to rhombic, 2.0–2.6 mm long, short-clawed. Fruits suborbiculoid to quadrate, with a winged, dentate keel on the back, usually umbonate at the base, 1.5–3.0 mm long (excluding the beak) and broad, with a marginal beak to 1 mm long, brown.

COMMON NAME: Pondweed.

HABITAT: Ponds, lakes, and streams.

RANGE: Quebec to British Columbia, south to northern California, Nebraska, and Virginia.

ILLINOIS DISTRIBUTION: Apparently rare; known from five counties in extreme northern Illinois and from Menard County.

Potamogeton zosteriformis may be recognized from most other Illinois pondweeds by its flattened stem and many-nerved leaves. Some previous workers have called this species *P. zosterifolius,* but this latter binomial was used for a different species by Schuman in 1801.

7. Potamogeton foliosus Raf. Med. Rep. II. 5:354. 1808. *Fig. 28.*

Potamogeton niagarensis Tuckerm. Am. Journ. Sci. II. 7:354. 1849.

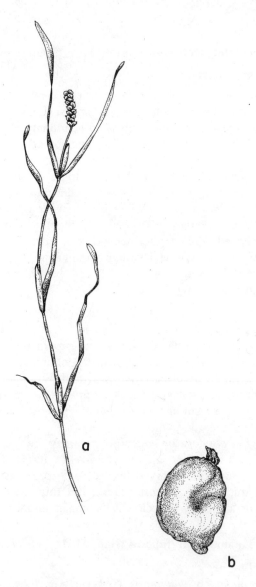

27. *Potamogeton zosteriformis* (Pondweed).   *a*. Habit, X½.   *b*. Fruit, X5.

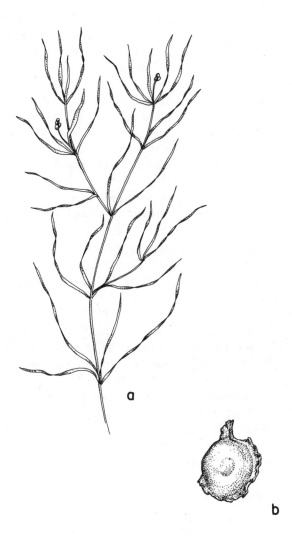

28. Potamogeton foliosus (Pondweed). *a.* Habit, X⅗. *b.* Fruit, 3⅘.

Potamogeton pauciflorus var. *niagarensis* (Tuckerm.) Gray, Man. Bot. 435. 1856.

Potamogeton foliosus var. *niagarensis* (Tuckerm.) Morong, Mem. Torrey Club 3:39. 1893.

Potamogeton foliosus var. *genuinus* Fern. Mem. Acad. Arts & Sci. 17:43. 1932.

Potamogeton foliosus var. *macellus* Fern. Mem. Acad. Arts & Sci. 17:46. 1932.

Rhizomes filiform, fibrous, rooting at the nodes. Stem filiform, flattened, simple to much branched. Leaves uniform, submersed, elongate, linear, subacute at apex, subcuneate at base, 2–14 cm long, up to 2.5 mm broad, green or bronze, 1- to 5-nerved; stipules connate at first to form tubular sheaths up to 17 mm long, at length splitting and falling away. Spikes continuous, subcapitate, cylindric, composed of 2–3 whorls of flowers, in fruit up to 5 mm long, 2–5 mm thick; flowers sessile; sepaloid connectives suborbicular to rhombic, 0.6–1.0 mm long, short-clawed, green or brownish. Fruits obovoid, compressed, dentate and strongly one-keeled on the back, 1.8–2.5 mm long (excluding the beak), nearly as broad, with a short beak less than 0.5 mm long, greenish-brown; $2n = 28$ (Stern, 1961).

COMMON NAME: Pondweed.

HABITAT: Ponds, lakes, rivers, and streams.

RANGE: Nova Scotia to British Columbia, south into Mexico; West Indies.

ILLINOIS DISTRIBUTION: Locally throughout the state. This is one of the more frequently encountered species of pondweeds in Illinois. It is distinguished by its few-flowered, subcapitate spikes and its tiny fruits which are dentate along the back.

Tuckerman's *P. niagarensis* is identical to *P. foliosus*. The slender form known as var. *macellus* grades into the typical material so well that it is not possible to divide the species into recognizable sub-units. A few early workers followed Pursh in calling this species *P. pauciflorus,* an illegitimate name.

8. Potamogeton friesii Rupr. Beitr. Pfl. Russ. Reich. 4:43. 1845. *Fig. 29.*

Potamogeton pusillus var. *major* Fries, Nov. Fl. Suec. 48. 1828.

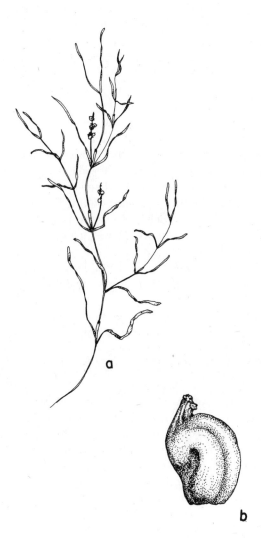

29. *Potamogeton friesii* (Pondweed). *a*. Habit, X⅔. *b*. Fruit, X5.

Potamogeton mucronatus Schrad. ex Reich. Ic. Fl. Germ. Helv. 7:15. 1845, in synon.

Rhizome slender, fibrous. Stem simple to sparsely branched, flattened. Leaves uniform, submersed, translucent, filiform to linear, obtuse to subacute to acute at apex, cuneate at base, 2.0–10.4 cm long, 0.7–3.5 mm broad, 5- to 7-nerved, with two glands at base; stipules at first connate forming a tubular sheath 7–16 mm long, whitish, fibrous. Spikes interrupted, cylindric, composed of 3–4 whorls of flowers, in fruit up to 1.5 cm long; flowers sessile; sepaloid connectives suborbicular to rhombic, 1.5–2.5 mm long, short-clawed. Fruits obovoid, rounded and weakly 3-keeled on the back, 2–3 mm long (excluding the beak), with a beak less than 1 mm long, brownish.

COMMON NAME: Pondweed.

HABITAT: Ponds and streams.

RANGE: Newfoundland to British Columbia, south to California, Illinois, and Virginia.

ILLINOIS DISTRIBUTION: Confined to two extreme northeastern counties (Cook: South Chicago, *Hill s.n.;* Lake: Lake Zurich, *Richardson 11511*). The report from Jackson County (Jones, *et al.*, 1955) could not be verified.

Potamogeton friesii is distinguished from the closely related *P. strictifolius* and *P. pusillus* by its translucent leaves which are 5- to 7-nerved.

Patterson and Brendel referred this species to *P. compressus* of Linnaeus, which it clearly is not.

9. **Potamogeton strictifolius** Benn. Journ. Bot. 40:148. 1902.

Fig. 30.

Rootstock slender. Stems filiform, compressed, simple to sparsely branched, the branchlets rigid. Leaves firm, uniform, submersed, linear, obtuse and sometimes mucronate, to 2.5 mm broad, 3-nerved, the margins more or less revolute; stipules connate at first, whitish, chartaceous, splitting and becoming fibrous at maturity, to 2 cm long. Spikes interrupted, cylindric, composed of 3–4 whorls of flowers, in fruit 0.6–1.5 cm long; flowers sessile; sepaloid connectives suborbicular, (1.3–) 1.5–1.8 mm long, short-clawed. Fruits obovoid to ovoid, rounded on the back, 2–3 mm long (excluding the beak), 1.5–2.0 mm broad, the marginal beak less than 1 mm long.

30. *Potamogeton strictifolius* (Pondweed). *a.* Habit, X⅛. *b.* Fruit, X3¾.

COMMON NAME: Pondweed.

HABITAT: Lakes.

RANGE: Quebec to Mackenzie, south to Utah, northern Illinois, and New York.

ILLINOIS DISTRIBUTION: Known only from two collections by Agnes Chase in 1900 and 1901 from Wolf Lake, Cook County.

This pondweed is similar to *P. pusillus* from which it differs by its fibrous stipules and slightly larger fruits.

10. **Potamogeton pusillus** L. Sp. Pl. 127. 1753. *Fig. 31.*

Potamogeton panormitanus var. *minor* Biv. Figlio Andrea 6. 1838.

Potamogeton panormitanus var. *major* Fischer, Ber. Bayer Bot. Gesells. 11:109. 1907.

Potamogeton pusillus var. *minor* (Biv.) Fern. & Schub. Rhodora 50:154. 1948.

Rootstock very slender. Stem filiform, compressed, much branched. Leaves uniform, submersed, linear, obtuse to acute at the apex, tapering to the base, 3–21 mm long, 0.5–2.0 mm broad, 3-nerved, usually bearing two translucent glands near the base; stipules connate to about the middle, membranous, splitting at maturity but not becoming fibrous, to 1.7 cm long. Spikes interrupted, subcylindric, composed of 2–5 whorls of flowers, in fruit 4–15 mm long; flowers sessile; sepaloid connectives suborbicular, 1.3–1.7 mm long, short-clawed. Fruits obovoid to reniform, smooth and very slightly one-keeled on the back, 2.0–2.8 mm long (excluding the beak), 1.0–1.8 mm broad, the beak about 0.5 mm long, greenish-brown; n = 26 (Palmgren, 1939).

COMMON NAME: Pondweed.

HABITAT: Ponds, lakes, and streams.

RANGE: Quebec to British Columbia, south to Texas and Alabama; Mexico; Europe.

ILLINOIS DISTRIBUTION: Local throughout the state.

Potamogeton pusillus is distinguished from *P. pectinatus* by its non-septate, shorter leaves, its much interrupted spikes, and its smaller fruits. From *P. friesii* and *P. strictifolius* it differs by its non-fibrous stipules. Size of the individual plants and their parts is variable.

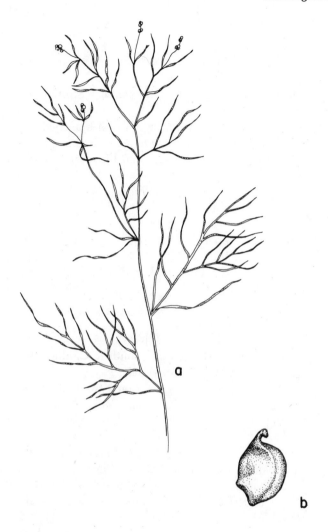

31. *Potamogeton pusillus* (Pondweed). *a.* Habit, X⅛. *b.* Fruit, X3¾.

Those specimens with more robust features have been segregated as var. *major,* while those with leaves less than 1 mm broad have been called var. *minor.* All of these variants occur in Illinois, but gradation is such that it is not feasible to recognize these lesser taxa.

11. Potamogeton berchtoldii Fieber, Oekon. Fl. Boeh. II. 1: 277. 1838. *Fig. 32.*

Potamogeton pusillus var. *tenuissimus* Mert. & Koch in Roehling, Deutsch. Fl. 1:857. 1823.
Potamogeton berchtoldii var. *tenuissimus* (Mert. & Koch) Fern. Rhodora 42:246. 1940.

Rootstock slender. Stem filiform, little to much branched. Leaves uniform, submersed, linear, subacute to acute at the apex, tapering to the base, to 5 cm long, 0.5–1.5 mm broad, 3-nerved, with a pair of translucent glands at the base; stipules free, but the margins inrolled, obtuse, to 1.5 cm long. Spikes continuous or slightly interrupted, subgloboid, composed of 1–3 whorls of flowers, in fruit 2–8 mm long, nearly as thick; flowers sessile; sepaloid connectives suborbicular, about 1.5 mm long. Fruits obovoid, rugulose, rounded on the back, 2.0–2.5 mm long (excluding the beak), 1.2–2.0 mm broad, with a beak about 0.5 mm long, dark greenish-brown.

COMMON NAME: Pondweed.
HABITAT: Wet ditches (in Illinois).
RANGE: Greenland to Alaska, south to California, Louisiana, and Virginia; Europe; Asia.
ILLINOIS DISTRIBUTION: Known only from collections by Hill from two localities from Cook County (South Chicago and Englewood) in 1875 and 1880.

Fernald (1950) recognizes six varieties of *P. berchtoldii* in the eastern United States, based on leaf apices, leaf widths, and leaf venations. Illinois material is referable to var. *tenuissimus,* but the dividing lines between the varieties are so uncertain that recognition of these lesser taxa seems undesirable.

Potamogeton berchtoldii is the only member of subsection Pusilli in Illinois in which the stipules are free.

12. Potamogeton vaseyi Robbins in Gray, Man. Bot. 485. 1867. *Fig. 33.*

Rootstocks slender, fibrous. Stem filiform, much branched, those with flowers and fruits bearing expanded leaves, those without flowers and fruits bearing unexpanded leaves. Leaves dimorphic; floating leaves elliptic to oval, obtuse at the apex, tapering to the base, 0.6–1.5 cm long, 3–7 mm broad, 5- to 9-nerved, the

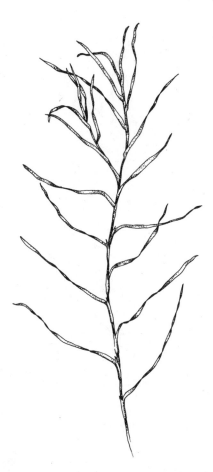

32. Potamogeton berchtoldii (Pondweed). X¼.

margins entire, the petiole up to twice as long as the blade;
submersed leaves linear-filiform, delicate, mucronulate at the
apex, tapering to the sessile base, 3.0–9.5 cm long, up to 0.5
mm wide; stipules slender, delicate, to 1.2 cm long. Spikes interrupted, cylindric, composed of 3–8 whorls of flowers, in fruit
6–9 mm long; flowers sessile; sepaloid connectives suborbicular,
about 1 mm long, short-clawed. Fruits suborbiculoid, flattened,
strongly keeled on the back, 0.9–2.0 mm in diameter, with a
marginal recurved beak about 0.5 mm long, greenish; $2n = 28$
(Harada, 1942).

33. *Potamogeton vaseyi* (Pondweed). *a.* Habit, X⅕. *b.* Fruit, X3¾.

a

b

COMMON NAME: Pondweed.

HABITAT: Lakes and ponds.

RANGE: Quebec to Minnesota, south to Iowa and New York.

ILLINOIS DISTRIBUTION: Known only in Illinois from the type locality at Ringwood, McHenry County, collected by George Vasey, and from Grundy County, where it was collected by Rowlatt in 1968.

Sterile specimens have unexpanded leaves, while fruiting specimens have dilated leaves.

I have been unable to check on the validity of the report of this species from Henry County (Dobbs, 1963).

13. **Potamogeton natans** L. Sp. Pl. 126. 1753. *Fig. 34.*

Rootstock thin, fibrous, white with reddish spots. Stem slender, simple to rarely branched, with ridges. Leaves dimorphic, coriaceous; floating leaves lanceolate to elliptic to ovate, obtuse at the apex, rounded to cordate at the base, 4–15 cm long, 2–6 cm broad, 13- to 37-nerved, with about one-third of the nerves more prominent, with petioles 3–15 cm long; submersed leaves long-linear, obtuse at the apex, tapering to the sessile base, 10–20 cm long, 0.5–2.0 mm broad, obscurely 3- to 5-nerved; stipules clasping, persistent, fibrous, whitish, linear to lanceolate, 3–11 cm long, 2-keeled. Spikes continuous, cylindrical, composed of 8–14 whorls of flowers, in fruit 2.5–6.0 cm long, 0.9–1.2 cm thick; flowers sessile or nearly so; sepaloid connectives suborbicular, 1.6–2.8 mm broad, short-clawed. Fruits obovoid, smooth and without a keel on the back, 3–5 mm long (excluding the beak), 2.0–3.5 mm broad, with a broad beak less than 1 mm long; n = 26 (Palmgren, 1939), 21 (Stern, 1961).

COMMON NAME: Pondweed.

HABITAT: Lakes and streams.

RANGE: Greenland to Alaska, south to California, Illinois, and North Carolina.

ILLINOIS DISTRIBUTION: Scattered throughout the state, except in the extreme southern counties where it is as yet uncollected.

Potamogeton natans is distinguished by its characteristic punctations on the rootstock and lower stems and its strongly developed cordate floating leaves.

Potamogeton nodosus, a similar species, has a distinct keel on the fruit, lacks the punctations on the rootstocks, and lacks the cordate floating leaves.

14. **Potamogeton diversifolius** Raf. Med. Repos. II 5:354. 1808. *Fig. 35.*

Rootstocks slender, fibrous. Stems profusely branched, occasionally simple, slender. Leaves dimorphic; floating leaves elliptic, oval, or narrowly obovate, broadly rounded at both ends, 1.2–3.5 cm long, 0.6–1.3 cm broad, the compressed petiole 1.7–2.0 cm long, the stipules to 3 cm long, becoming fibrous; submersed leaves linear, obtuse to subacute at the apex, narrowed to the base, to 45 mm long, to 1.5 mm broad, sessile, the stipules 2–15 mm long, sheathing for one-half of their length. Spikes

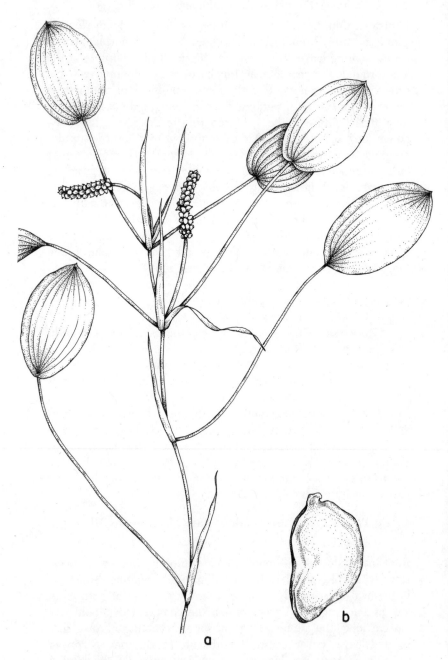

34. Potamogeton natans (Pondweed). *a.* Habit, X¼. *b.* Fruit, X3¾.

35. *Potamogeton diversifolius* (Pondweed).   *a.* Habit, X⅖.   *b.* Fruit, X10.

continuous, subgloboid to elongate, 0.6–1.2 (–2.0) cm long, those in the axils of the submersed leaves subgloboid, those in the axils of the floating leaves elongate; flowers sessile; sepaloid connectives suborbicular, 0.7–1.0 mm long, short-clawed. Fruits suborbiculoid to reniform, smooth and with one strong and two weak keels on the back, 1.0–1.5 mm long (excluding the beak), 1.0–1.5 mm broad, with a minute beak, green to greenish-brown.

COMMON NAME: Pondweed.

HABITAT: Quiet waters.

RANGE: New Jersey to Minnesota to Oregon, south to Texas, Illinois, and West Virginia, Georgia and Florida; Mexico.

ILLINOIS DISTRIBUTION: Throughout the state; seemingly more common in the southern tip of Illinois.

Potamogeton diversifolius is the only Illinois representative of subsection Hybridi, characterized by the dimorphic spikes.

Considerable misunderstanding among Illinois workers has existed in the past. Specimens from Illinois which actually are *P. diversifolius* have been reported as *P. hybridus* by Lapham (1857), Patterson (1874, 1876), and Brendel (1887), as *P. spirillus* by Patterson (1876) and Brendel (1887), as *P. dimorphus* by Pepoon (1927), and as *P. capillaceus* by Kibbe (1952).

15. Potamogeton epihydrus Raf. Med. Repos. II 5:354. 1808.

Fig. 36.

Potamogeton claytonii Tuckerm. Am. Journ. Sci. 45:38. 1843.
Potamogeton epihydrus var. *typicus* Fern. Mem. Am. Acad. Arts & Sci. 17:114. 1932.

Rootstocks extensively creeping, slender. Stems compressed, simple or little branched. Leaves dimorphic; floating leaves usually opposite, elliptic to oblong-lanceolate, obtuse and sometimes cuspidate, to 8 cm long, to 3.5 cm broad, 19- to 41-nerved, coriaceous, with compressed petioles, with subcoriaceous, attenuated stipules; submersed leaves linear-elongate, to 20 cm long, 0.5–1.0 cm broad, 7- to 13-nerved, the free stipules membranous, obtuse, 2.0–4.5 cm long; transitional leaves usually present. Spikes continuous, cylindrical, to 4 cm long; flowers sessile; sepaloid connectives flabellate, 1.5–3.0 mm long, short-clawed. Fruits rhombic-obovoid, laterally compressed, 3-keeled, pitted, 3.0–4.5 mm long (excluding the beak), 3.0–3.5 mm broad, with a merely tooth-like beak.

36. Potamogeton epihydrus (Pondweed). *a.* Habit, X⅛. *b.* Fruit, X3.

COMMON NAME: Pondweed.

HABITAT: Quiet ponds and lakes.

RANGE: Quebec to British Columbia, south to California, Illinois, West Virginia, and New Jersey.

ILLINOIS DISTRIBUTION: Collected from three northern counties, all before 1900.

The distinguishing characters of this species are the compressed stem, the submersed leaves 5–10 mm broad, and the tooth-like beak of the achene.

This species, the only Illinois representative of subsection Nuttalliana, may now be extinct in the Illinois flora.

16. Potamogeton amplifolius Tuckerm. Am. Journ. Sci. 6: 225. 1848. *Fig. 37.*

Rootstocks thick, to 4 mm in diameter, whitish or reddish, with obtuse, black scales. Stems terete, somewhat less in diameter than the rootstocks, simple or becoming somewhat branched at maturity. Leaves trimorphic; floating leaves ovate to elliptic, obtuse and sometimes mucronate at the apex, cuneate or rounded at the base, coriaceous, 5–10 cm long, 2.5–5.0 cm broad, 21- to 51-nerved, with about one-fourth of the nerves more conspicuous than the others, with a petiole to 20 cm long, with the stipule persistent, fibrous, 2-keeled, to 20 cm long, 30- to 40-nerved; submersed leaves on lower part of stem lanceolate, obtuse to subacute at the apex, to 20 cm long, 2.5–7.5 cm broad, with a petiole to 6 cm long; submersed leaves on the upper part of the stem broadly lanceolate to ovate, obtuse to subacute at the apex, to 20 cm long, to 7.5 cm broad, 23- to 37-nerved, with the petioles to 6 cm long, with the stipules persistent, fibrous, obscurely keeled, to 11 cm long. Spike continuous, composed of 9–16 whorls of flowers, in fruit 4–8 cm long, 1.0–1.5 cm thick; flowers sessile; sepaloid connectives suborbicular, 2–4 mm long, clawed. Fruit obovoid, rounded and usually prominently keeled on the back, 3.5–5.0 mm long (excluding the beak), 2.5–4.0 mm broad, the usually prominent beak about 1 mm long; $n = 26$ (Stern, 1961).

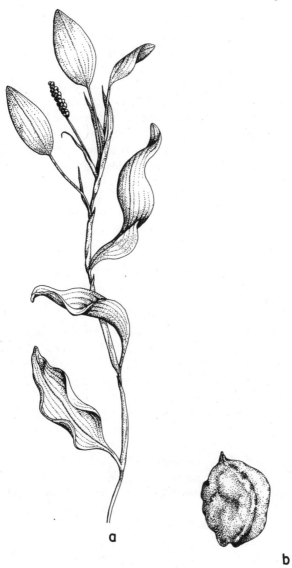

a

b

37. Potamogeton amplifolius (Pondweed). *a.* Habit, X⅛. *b.* Fruit, X2½.

COMMON NAME: Pondweed.

HABITAT: Lakes and roadside ditches.

RANGE: Newfoundland to British Columbia, south to California, Oklahoma, and Georgia.

ILLINOIS DISTRIBUTION: Not common; restricted to the northern one-half of Illinois; attributed to Wabash County by Jones (1963), but not verified.

The diagnostic features of *P. amplifolius* are its robust stature, its large submersed leaves just below the surface of the water, and its floating leaves with numerous nerves.

It differs from *P. pulcher,* the other Illinois member of subsection Amplifolii, by its larger spikes, more persistent stipules, cuneate achenes, and its veins of different degrees of prominence.

17. Potamogeton pulcher Tuckerm. Am. Journ. Sci. 45:38. 1843. *Fig. 38.*

Rootstocks slender, up to 1 mm in diameter, usually red-spotted. Stems terete, to 2.5 mm in diameter, simple, black-punctate. Leaves trimorphic; floating leaves ovate to orbicular, obtuse and sometimes bluntly mucronate at the apex, cordate or rounded at the base, to 7 (–11) cm long, 1.5–5.5 (–8.5) cm broad, 19- to 35-nerved, coriaceous, with the petiole to 18 cm long, usually much shorter, with the stipules persistent, obtuse to subacute, 2–5 cm long, two-keeled; submersed leaves to 18 cm long, to 3.5 cm broad, 9- to 21-nerved, with the petiole up to 1.5 cm long, with the stipules early decaying, lower leaves oblong, obtuse, subcoriaceous, upper leaves lanceolate to lancelinear, acute or subacute, translucent; transitional leaves present. Spikes continuous, cylindrical, composed of about 10 whorls of flowers, in fruit 2.0–3.5 cm long, about 1 cm thick; flowers sessile or nearly so; sepaloid connectives suborbicular, 1.5–4.0 mm long, clawed. Fruits obliquely ovoid, rounded on the back with one prominent and two rather obscure keels, 2–4 mm long (excluding the beak), 2.0–3.5 mm broad, with a prominent beak up to 0.8 mm long, light brown to olive-green.

COMMON NAME: Pondweed.

HABITAT: Shallow water.

RANGE: Nova Scotia to Minnesota, south to Texas, Ohio, and Maryland; Florida.

ILLINOIS DISTRIBUTION: Known from Menard and Mason counties (*Hall* in 1861) and Jackson County (Campbell Lake, *Bailey & Swayne 3109*). The St. Clair County record reported by Jones (1963) was not verified.

The conspicuously black-punctate leaves, the large cordate floating leaves, and the lanceolate submersed leaves abruptly tapering at base serve to distinguish this rare pondweed.

38. Potamogeton pulcher (Pondweed).   *a.* Habit, X⅛.   *b.* Fruit, X3¾.

Occasional plants in very shallow water have all the leaves floating.

18. Potamogeton nodosus Poir. Lam. Encycl. Meth. Bot. Suppl. 4:535. 1816. *Fig. 39.*

Potamogeton occidentalis Sieb. ex Cham. & Schlecht. Linnaea 2:224. 1827.

Potamogeton americanus Cham. & Schlecht. Linnaea 2:226. 1827.

Rootstocks thick, reddish-spotted. Stems terete, simple, to 2 mm in diameter. Leaves dimorphic; floating leaves elliptic, obtuse to subacute at the apex, cuneate or rounded at the base, 9- to 21-nerved, coriaceous, 3–15 cm long, 1.5–5.2 cm broad; submersed leaves linear-lanceolate to elliptic, subacute at the apex, tapering to the base, denticulate (at least when young), thin, to 20 cm long, to 3.5 cm broad, 7- to 15-nerved, with the compressed petiole 2–13 cm long; stipules linear, acute or obtuse, delicate, brownish, 2–9 cm long, usually two-keeled. Spike more or less continuous, cylindrical, composed of 10–17 whorls of flowers, in fruit 3–10 cm long, about 1 cm thick; flowers sessile; sepaloid connectives suborbicular, 1.4–2.6 mm long, short-clawed. Fruits obovoid, smooth and prominently one-keeled on the back, 2.0–4.0 (–4.3) mm long (excluding the beak), 2.5–3.0 mm broad, with a short beak less than 1 mm long, brownish-red.

COMMON NAME: Pondweed.

HABITAT: Ponds and streams.

RANGE: New Brunswick to British Columbia, south to Texas and Alabama; Mexico.

ILLINOIS DISTRIBUTION: Widely distributed throughout the state.

Potamogeton nodosus is recognized by its cuneate floating leaves, its narrowly lanceolate submersed leaves tapering to either end, and its rather large, prominently keeled achenes.

Considerable variation is exhibited both by the floating and the submersed leaves. Length and compactness of the spike are also variable.

Many Illinois workers, questioning the identity of Poiret's *P. nodosus,* have called this species *P. americanus.* Tuckerman's *P. lonchites,* attributed to Illinois by Schneck (1876), Patterson (1876), and Brendel (1887), is based in part on *P. nodosus.*

39. *Potamogeton nodosus* (Pondweed). *a.* Habit, X⅛. *b.* Fruit, X3¾.

19. **Potamogeton illinoensis** Morong, Bot. Gaz. 5:50. 1880.
Fig. 40.

Rootstocks stout. Stems slender to rather stout, up to 5 mm in diameter, simple to little branched. Leaves dimorphic; floating leaves (often absent) elliptic to ovate, obtuse and mucronate, rounded or tapering to the base, denticulate, to 15 (–20) cm long, 2–5 (–6.5) cm broad, 13- to 29-nerved, with a petiole shorter than the blade; submersed leaves linear to lanceolate to narrowly ovate, acute and usually mucronate at the apex, rounded or tapering to the base, denticulate, to 20 cm long, to 4.5 cm broad, with a petiole somewhat compressed, up to 3.5 cm long; stipules obtuse, to 5 (–8) cm long, two-keeled. Spike compact, continuous, composed of 8–15 whorls of flowers, in fruit to 7 cm long, about 1 cm thick; flowers sessile; sepaloid connectives suborbicular, 1.5–3.0 mm long, short-clawed. Fruits ovoid to suborbiculoid, with one prominent and two usually obscure keels, 2.5–3.5 mm long, 2–3 mm broad, with the beak about 0.5 mm long, brown; $n = 52$ (Stern, 1961).

COMMON NAME: Pondweed.

HABITAT: Ponds, lakes, and streams.

RANGE: Quebec to British Columbia, south to California, Texas, and Florida.

ILLINOIS DISTRIBUTION: Occasional in the northern half of Illinois; also Macoupin, Pope, and Wabash counties. The type was collected by Patterson near Oquawka, Henderson County.

Potamogeton illinoensis is similar to *P. gramineus*, but differs by its little-branched stems, its prominently keeled stipules, and its generally larger spikes, leaves, and sepaloid connectives.

This is a highly complex species. Some sterile specimens resemble the European *P. lucens*, but they lack the tendency for the lower leaves to have the blades reduced at the apex so that the midribs form a cusp at apex.

40. *Potamogeton illinoensis* (Pondweed). *a.* Habit, X⅙. *b.* Fruit, X3¾.

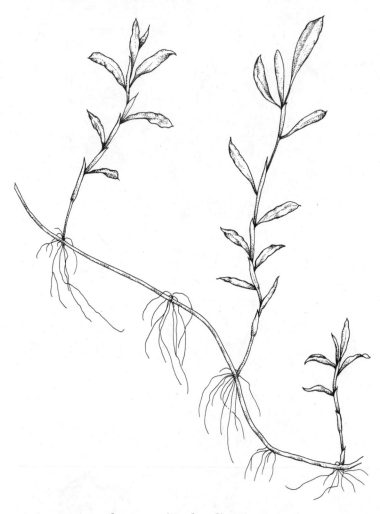

41. Potamogeton × hagstromii (Pondweed). X¼.

20. Potamogeton × hagstromii Benn. Trans. & Proc. Bot. Soc. Edinburgh 29:51. 1924. *Fig. 41.*

Potamogeton epihydrus × angustifolius Benn. Journ. Bot. 42: 71. 1904.
Potamogeton claytonii × zizii Graebn. in Engler, Pflanzenr. 4(11):133. 1907.
Potamogeton subdentatus × petiolatus Hagstr. Crit. Res. Pot. 201. 1916.

Rootstocks moderately thick. Stems simple, 1.0–2.5 cm in diameter. Leaves dimorphic; floating leaves lanceolate, acute at the apex, tapering to base, to 3.5 cm long, to 0.7 cm broad, with the petiole to 1 cm long; submersed leaves linear-lanceolate, acute at the apex, tapering to the sessile base, denticulate, to 7 cm long, to 1.5 cm broad, 11-nerved; stipules elongate, pellucid, tenuous, with prominent nerves. Spikes more or less continuous; flowers sessile; sepaloid connectives suborbicular, 1.0–2.5 mm long, short-clawed. Immature fruits suborbiculoid.

COMMON NAME: Pondweed.
HABITAT: Lakes.
RANGE: Scattered throughout North America.
ILLINOIS DISTRIBUTION: Known only from Cook County (Wolf Lake, *A. Chase 1713;* locality unknown, *Sherff s.n.*). Both of these specimens are in the herbarium of the Field Museum of Natural History.
This is the reputed hybrid between *P. richardsonii* and *P. gramineus*. Because of the dimorphic leaves, the resemblance is more to *P. gramineus*.

This hybrid is remarkable for the stipules which embrace the entire internode from leaf to leaf in the upper part of the stem.

21. **Potamogeton gramineus** L. Sp. Pl. 1:127. 1753. *Fig. 42.*

Potamogeton gramineus var. *graminifolius* Fries, Nov. Fl. Suec. 36. 1828.

Potamogeton heterophyllus f. *graminifolius* (Fries) Morong, Mem. Torrey Club 3(2):24. 1893.

Potamogeton gramineus var. *typicus* Ogden, Rhodora 45:143. 1943.

Rootstocks thin or thickish, frequently red-spotted. Stems much branched, to 1 mm in diameter. Leaves dimorphic; floating leaves elliptic to ovate, obtuse and often mucronate at the apex, tapering or rounded at the base, coriaceous, to 7 cm long, 1–3 cm broad, 13- to 23-nerved, with a petiole to 15 cm long; submersed leaves narrowly elliptic to oblanceolate, acute at the apex, tapering to the base, sessile, denticulate, to 7 cm long, 2–8 mm broad; stipules persistent, obtuse at the apex, to 3 cm long, 1–5 mm broad, obscurely two-keeled. Spike generally compact, comprised of 5–10 whorls of flowers, in fruit 1–3 cm long, 5–10 mm broad; flowers sessile or on pedicels up to 0.5 mm long; sepaloid connectives suborbicular, 0.7–2.5 mm long and broad, short-clawed. Fruit obovoid, prominently keeled, 1.5–3.0 mm long (excluding the beak), 1.5–2.5 mm broad, with a short curved beak, greenish; n = 26 (Palmgren, 1939).

COMMON NAME: Pondweed.

HABITAT: Lakes, ponds, and streams.

RANGE: Newfoundland to Alaska, south to California, Arizona, northern Illinois, Maryland, and New York; Greenland; Europe; Asia.

ILLINOIS DISTRIBUTION: Rare; recorded only from Cook (Hill *s.n.*) and Wabash (Richardson *s.n.*) counties. Other records reported by Winterringer and Evers (1960) are misidentifications.

The much branched stems and the numerous small leaves are diagnostic for this species.

Potamogeton gramineus is extremely variable in leaf shape and number of nerves per leaf. Only the typical variety has been collected in Illinois, although var. *maximus* Morong with larger submersed leaves and 7–11 nerves per leaf and var. *myriophyllus* Robbins with narrower submersed leaves and 3 nerves per leaf could occur in Illinois.

22. **Potamogeton** × **spathulaeformis** (Robbins) Morong, Mem. Torrey Club 3(2):26. 1893. *Fig. 43.*

Potamogeton gramineus var. *spathulaeformis* Robbins in Gray, Man. 487. 1867.

Rootstocks extensively creeping. Stems much branched, sometimes attaining a length of nearly one meter. Leaves dimorphic; floating leaves elliptic to obovate, acute at the apex, tapering to the base, to 5.5 cm long, 0.5–2.5 cm broad, 13- to 23-nerved, with a petiole to 5 cm long; submersed leaves reduced to phyllodia or linear-lanceolate to oblong, cuspidate at the apex, tapering to the base, pellucid, denticulate, to 5 cm long, to 1 cm broad, 5- to 13-nerved, sessile or on very short petioles; stipules obtuse and cucullate at the apex, eventually splitting, obscurely keeled. Spikes compact, densely flowered; flowers sessile; sepaloid connectives suborbicular, 1–3 mm long, short-clawed. Fruits obovoid or suborbiculoid, obscurely three-keeled, 1.5–3.0 mm long (excluding the beak), 1.5–2.5 mm broad, with a short curved beak.

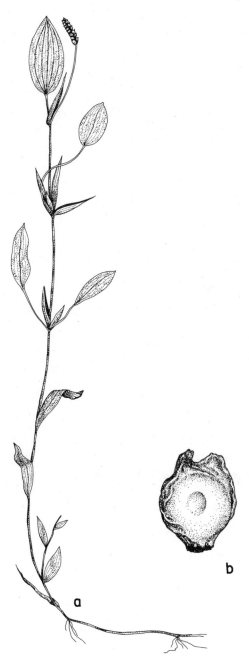

42. Potamogeton gramineus (Pondweed). *a.* Habit, X¼. *b.* Fruit, X6¼.

43. Potamogeton × *spathulaeformis* (Pondweed). X¼.

COMMON NAME: Pondweed.

HABITAT: Ponds and streams.

RANGE: Ontario and Vermont to Idaho, south to Arizona, Iowa, and New York.

ILLINOIS DISTRIBUTION: A single collection from Wabash County is known (near Mt. Carmel, *J. Schneck s.n.*). The parents of this hybrid are reputed to be *P. illinoensis* and *P. gramineus.* The branched stems, size of leaves, and achene characters are very similar to *P. gramineus,* while the tendency for the submersed leaves to become petiolate is like *P. illinoensis.*

23. Potamogeton × rectifolius Benn. Journ. Bot. 40:147. 1902.
Fig. 44.

Potamogeton nodosus × richardsonii Hagstr. Crit. Res. Pot. 148. 1916.

Rootstocks creeping, thick. Stems branched, to 6 cm long. Leaves dimorphic; floating leaves oblong- to linear-lanceolate, obtuse at the apex, tapering at the base, coriaceous, to 10 cm long, to 4 cm broad, 12- to 14-nerved, with a petiole 2–4 cm long; submersed leaves linear, acute to subacute at the apex, tapering at the base, pellucid, denticulate, to 15 cm long, to 3 cm broad, 10- to 12-nerved, sessile or with a petiole to 1 cm long; stipules acuminate, to 2.5 cm long. Spikes more or less continuous, sparsely flowered; flowers sessile; sepaloid connectives orbicular, 1.5–2.5 mm in diameter. Fruits unknown.

COMMON NAME: Pondweed.

HABITAT: Ditches, in 1–3 feet of water.

RANGE: Illinois; Oregon.

ILLINOIS DISTRIBUTION: Known only from the type locality in Cook County (Stony Island, *Hill 71.1900*, the type; *Hill 179.1901; Hill 191.1902; A. Chase 1477; A. Chase 1994*).

This plant is a hybrid between *P. nodosus* and *P. richardsonii*. It is remarkable that the rare *P. richardsonii*, known only from three Illinois counties, is one of the parents for two hybrid pondweeds in Illinois.

RUPPIACEÆ – DITCH GRASS FAMILY

Only the following genus comprises this family.

1. *Ruppia* L. – Ditch Grass

Aquatic herbs; leaves capillary, sheathing at the base; flowers bisexual, borne two on a spadix; perianth 0; stamens 2, sessile; carpels 4, free, 1-celled, each with a single ovule; fruit drupaceous, borne on a stipe (podogyne).

Some authors prefer to group *Ruppia* with *Potamogeton* in the Potamogetonaceae, others place it with *Zannichellia*, while still others put it near *Najas*. There seem to be enough fundamental differences to merit its segregation into a separate family.

Only the following taxon occurs in Illinois.

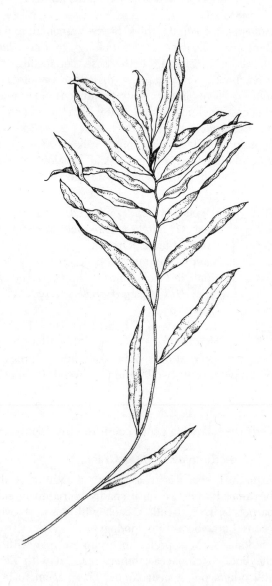

44. *Potamogeton × rectifolius* (Pondweed). X⅛.

1. **Ruppia maritima** L. var. **rostrata** Agardh, Physiog. Sallsk. Arbetr. 27. 1823. *Fig. 45.*

Ruppia rostellata Koch ex Reichenb. Ic. Pl. Crit. 2:66. 1824.

Slender herbs, submerged; stems filiform, branched; leaves alternate, capillary, up to 10 cm long, less than 0.5 mm wide, with membranous basal sheaths up to 1 cm long; bisexual flowers two per spadix, inclosed at anthesis by the sheathing leafbase; perianth none; stamens 2, each with 2 large locules; peduncles 1.5–3.0 mm long; carpels 4, free, each with a single suspended ovule; drupes slenderly and excentrically beaked, 1.5–3.0 mm long, the podogyne 10–35 mm long.

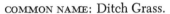

COMMON NAME: Ditch Grass.

HABITAT: Saline waters.

RANGE: Newfoundland to British Columbia, south to California; Oklahoma and Florida; rare inland; Mexico; Central America; West Indies; South America; Europe; Asia; Africa.

ILLINOIS DISTRIBUTION: Known from Vermilion, Lake, and Henry counties. First collected from Henry County on May 25, 1964, by *L. Roek s.n.*

Of the many varieties which are known of *Ruppia maritima,* variety *rostrata* is the most widespread. The distinguishing features of this variety are its excentrically beaked fruits and its long podogynes.

Fernald and Wiegand (1914) have given an account of the varieties which occur in North America.

ZANNICHELLIACEÆ – HORNED PONDWEED FAMILY

Anatomical evidence presented by Uhl (1947) supports the segregation of *Zannichellia* into a separate family.

Only the following genus comprises the family.

1. *Zannichellia* L. – Horned Pondweed

Monoecious perennial with rhizomes; leaves linear, not dilated at base, entire; flowers unisexual, reduced, axillary; perianth 0; stamen 1; pistils 2–4, free, 1-celled, each cell with a single ovule; nutlets in fascicles of 4, axillary, with a persistent, beak-like style.

Only the following species occurs in Illinois.

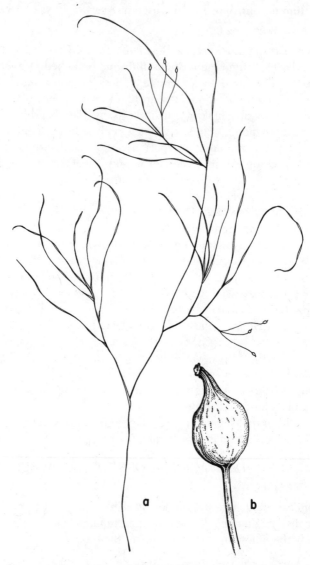

45. *Ruppia maritima* var. *rostrata* (Ditch Grass).   *a.* Habit, X¼.   *b.* Fruit, X15.

1. **Zannichellia palustris** L. Sp. Pl. 969. 1753. *Fig. 46.*

Zannichellia intermedia Torr. in Beck, Bot. N. & M. St. 385. 1833.

Slender monoecious perennials, submerged, with rhizomes; stems thread-like, sparsely branched, light green; leaves opposite, linear, entire, acute, dark green, 3–7 cm long, 0.4–0.7 mm wide, the base with clasping stipules, not dilated; unisexual flowers adjacent in same axil; perianth none; staminate flowers with a single stamen, the anther borne on a slender filament; pistillate flowers with 2–4 (usually 4) separate carpels, each with an orthotropous ovule, slender style, peltate stigma; nutlets usually in fascicles of 4, axillary, short-pedicellate, flattish, falcate, obliquely oblongoid, dentate on convex surface, brownish, the body 2.0–2.5 mm long, with persistent, beak-like style, 0.8–1.2 mm long.

COMMON NAME: Horned Pondweed.

HABITAT: Ditches, spring-fed streams, and ponds.

RANGE: Widespread throughout continental United States, north to Alaska; Europe; Asia.

ILLINOIS DISTRIBUTION: Occasional in the northern one-half of the state; absent elsewhere, except for Lake Murphysboro in Jackson County.

Although *Zannichellia palustris* resembles some other submerged aquatics, the beaked nutlets borne in fascicles of four are distinctive. Vegetatively, *Zannichellia* can be distinguished from *Najas* by its clasping stipules.

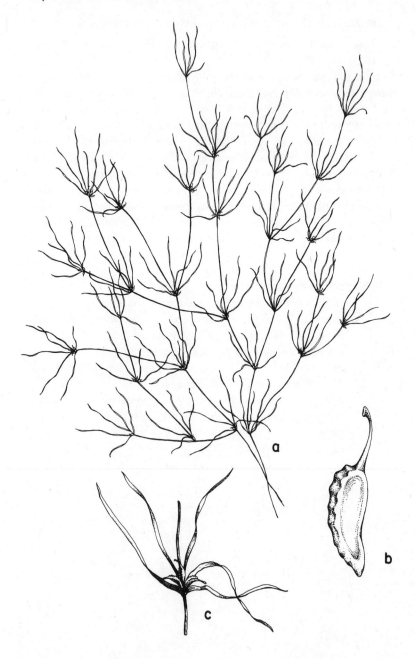

46. *Zannichellia palustris* (Horned Pondweed). *a.* Habit, X¼. *b.* Nutlet, X6¼. *c.* Cluster of fruits, X1.

Order Najadales

The following family comprises the order.

NAJADACEÆ – NAIAD FAMILY

Only the following genus comprises the family.

1. Najas L. – Naiad

Monoecious or dioecious, slender, branching aquatics, rooting from lower nodes; leaves linear, dilated at base, opposite, sessile, entire, serrulate, or coarsely toothed; flowers unisexual, reduced, usually solitary in leaf-sheath axils; perianth 0; staminate flowers borne near terminal nodes, a minute spathe enclosing a single stamen; pistillate flowers borne near middle and lower nodes, with a single 1-locular ovary and 1 ovule; stigmas 2–3, awl-shaped; achene oblongoid, with reticulate surface, enclosed in a closely adhering membranous epicarp.

For a detailed study of this genus in Illinois, see Fore and Mohlenbrock (1966) and Winterringer (1966).

KEY TO THE SPECIES OF Najas IN ILLINOIS

1. Leaves seemingly entire or serrulate; plants monoecious; achene 1.5–3.5 mm long.
    2. Leaves linear, 0.5–1.4 mm wide, with slightly dilated bases; marginal spinules microscopic.
        3. Achene lustrous, with 30–50 rows of obscure, minute, often square areolae; style and 2 stigmas 0.8–1.6 mm long_____ _____1. N. flexilis
        3. Achene dull, with 16–24 rows of distinct, large, hexagonal or rectangular areolae; style and 2–3 stigmas less than 1 mm long_____2. N. guadalupensis
    2. Leaves filiform, 0.1–0.3 mm wide, with abruptly dilated bases; marginal spinules macroscopic.
        4. Achene with roughened appearance, with 22–40 rows of longer than broad, rectangular areolae; style and 2–3 stigmas 0.8–1.2 mm long_____3. N. gracillima
        4. Achene with ribbed appearance, with 10–18 rows of broader than long, rectangular areolae; style and 2 stigmas 1.0–1.4 mm long_____4. N. minor
1. Leaves coarsely toothed; plants dioecious; achene 4.0–7.5 mm long _____5. N. marina

1. **Najas flexilis** (Willd.) Rostk. & Schmidt, Fl. Sed. 384. 1824. *Fig. 47.*

Caulinia flexilis Willd. Abh. Akad. Berlin 95. 1803.

Naias canadensis Michx. Fl. Bor. Am. 2:220. 1803.

Long and slender to closely tufted annuals; stems narrow, densely or sparsely branched, the terminal nodes crowded, the internodes dark green; leaves linear or sublanceolate, setaceous, light to dark green, 1–3 cm long, 0.5–1.4 mm wide, the apex usually acute and slightly recurved, gradually sloping to the base, ovate, slightly dilated, unlobed, each leaf-margin with 16–35 microscopic spinules less than 0.5 mm long; persistent style and 2 stigmas 0.8–1.6 mm long; epicarp yellowish to purplish; achene fusiform to slender, highly lustrous, yellowish to purplish-brown, 0.8–3.0 mm long, ⅛ as thick, marked with 30–50 longitudinal rows of obscure, usually squarish areolae; 2n = 12, 24 (Chase, 1947).

COMMON NAME: Naiad.

HABITAT: Shallow water of streams, lakes, and strip mine ponds.

RANGE: Newfoundland to Minnesota, south to Iowa and Virginia; also in the Pacific states; British Columbia; northwestern Europe.

ILLINOIS DISTRIBUTION: Occasional; throughout the state. All Illinois species of *Najas* except *N. marina* belong to subgenus Caulinia, a group characterized by spinulose basal sheaths, small fruits, and a monoecious condition. The four species of *Najas* which comprise this subgenus may be distinguished through a careful examination of the achenes. In *N. flexilis* and *N. guadalupensis*, the surface of the achene is covered by a pattern of nearly square or 6-sided areas, termed areolae; in *N. gracillima* and *N. minor*, these areolae are rectangular. The squarish areolae in *N. flexilis* are arranged in 30–50 rows, while in *N. guadalupensis*, there are 16–24 rows. The rectangular rows of areolae in *N. gracillima* are arranged vertically; in *N. minor*, they are arranged horizontally.

The evidence from the Illinois collections of *Najas* indicates that *N. flexilis* is the most widespread species.

2. **Najas guadalupensis** (Spreng.) Magnus, Beitr. Gatt. 8. 1870. *Fig. 48.*

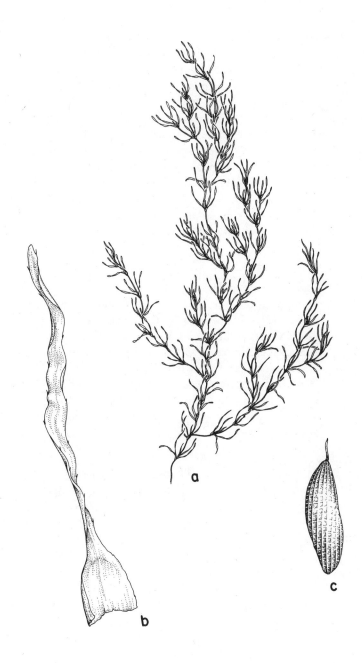

47. Najas flexilis (Naiad). *a.* Habit, X½. *b.* Leaf, X10. *c.* Fruit, X7½.

Caulinia guadalupensis Spreng. Syst. 1:20. 1825.

Naias flexilis var. *guadalupensis* A. Br. Journ. Bot. Brit. &
For. 2:276. 1864.

Usually long and slender annuals; stems slender, terete, rather
sparsely branched, the terminal nodes not very crowded, the
internodes dark green or purplish; leaves narrowly linear, setaceous, light green, 0.6–2.0 cm long, 0.5–1.2 mm wide, the apex
acute or obtuse but not recurved, gently sloping to the base,
slightly dilated, the shoulders rounded, unlobed, each leaf-
margin with 10–18 microscopic spinules less than 0.5 mm long;
persistent style and 2–3 stigmas less than 1 mm long; epicarp
purplish-brown; achene fusiform, ellipsoid, dull, light yellow
to brownish, 1.2–2.4 mm long, ⅓ as thick, marked with 16–24
rows of distinct areolae, hexagonal at first, generally becoming
more rectangular when dried at maturity; 2n = 12, 36, 42, 48,
54, 60 (Chase, 1947).

COMMON NAME: Naiad.

HABITAT: Along the shores of ponds and shallow lakes.

RANGE: Massachusetts to South Dakota, south to Texas;
Florida; also in the Pacific states; tropical America.

ILLINOIS DISTRIBUTION: Occasional in the southern counties; rare or absent elsewhere.

In addition to differences in achenes, *N. guadalupensis*
may be distinguished from the similar *N. flexilis* by its
fewer spinules along the leaf margins and by its
deeper green color.

3. **Najas gracillima** (A. Br.) Magnus, Beitr. 23. 1870. *Fig. 49.*

Naias indica var. *gracillima* A. Br. Journ. Bot. Brit. & For.
2:277. 1864.

Usually moderately compact annuals; stems quite slender, often
highly branched, most nodes somewhat crowded, the internodes
purplish; leaves filiform, linear, serrulate, light to dark green,
1.5–3.0 cm long, 0.1–0.3 mm wide, the apex acicular but not
recurved, the base scarious, abruptly dilated, auriculate with
4–7 antrorse, denticulate auricles, each leaf-margin with 8–20
macroscopic spinules more than 0.5 mm long; persistent style
and 2–3 stigmas 0.8–1.2 mm long; epicarp purplish; achene usually subfalcate to linear with oblique sides, dull to slightly
lustrous, yellowish to grayish, 2.5–3.5 mm long, ¼ as thick,

48. *Najas guadalupensis* (Naiad). *a.* Habit, X½. *b.* Leaf, X10. *c.* Fruit, X8¾.

marked with 22–40 vertical rows of longitudinal, rectangular areolae; 2n = 24, 36 (Chase, 1947).

COMMON NAME: Naiad.
HABITAT: Pools, ponds, and lakes with sandy and muddy substrata.
RANGE: Maine to Minnesota, south to Missouri and Virginia.
ILLINOIS DISTRIBUTION: Rare; known from three scattered counties.
This is the most slender and delicate species of *Najas* in Illinois. The tips of the leaves generally are not as recurved as in the other species.

4. **Najas minor** All. Fl. Pedem. 2:221. 1875. *Fig. 50.*

Slender to terminally compact annuals; stems slender, but slightly stouter than *N. gracillima,* moderately branched, the terminal nodes crowded, with lime-green internodes; leaves filiform, colored the same as the internodes, 2.0–3.5 cm long, 0.2–0.3 mm wide, the distal end spinescent, recurved, the sheath abruptly distended, truncate, serrulate, each leaf-margin with 8–16 barely macroscopic spinules more than 0.5 mm long; style and 2 stigmas 1.0–1.4 mm long; epicarp purplish; achene falcate, slender, oblique, 2.6–3.6 mm long, ¼ as wide, marked with 10–18 regular vertical ribs of broad, rectangular reticulations; 2n = 24 (Chase, 1947).

COMMON NAME: Naiad.
HABITAT: Shallow waters along lake shores.
RANGE: New York, West Virginia, Alabama, Tennessee, and Illinois; Asia; Europe.
ILLINOIS DISTRIBUTION: Rather common; restricted to the southern two-thirds of the state.
This species was first discovered in Illinois in 1963 from Lake Murphysboro. A native of the Old World, it is infrequently established in the United States. The Illinois stations are the farthest west in this country.
The horizontal areolae on the achene give the achene a ribbed appearance under magnification.

5. **Najas marina** L. Sp. Pl. 1015. 1753. *Fig. 51.*

Rather slender, delicate, dioecious annuals; stems branching, to

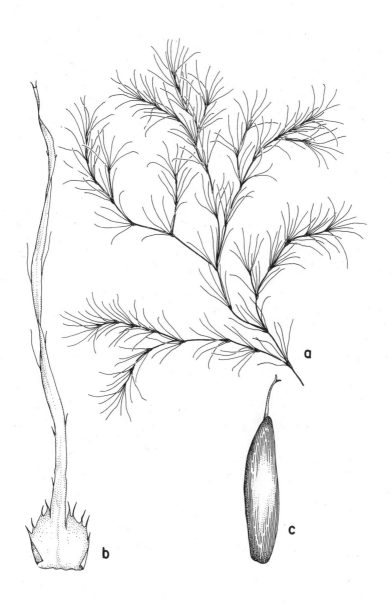

49. *Najas gracillima* (Naiad). *a.* Habit, X½. *b.* Leaf, X10. *c.* Fruit, X7½.

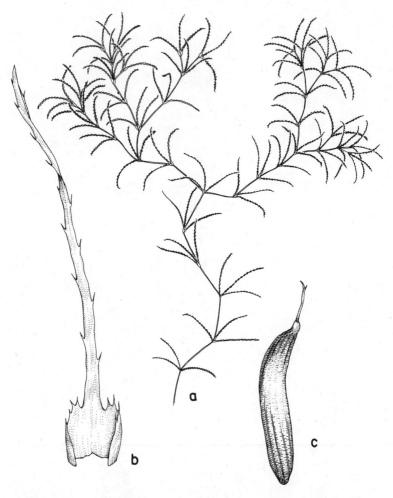

50. *Najas minor* (Naiad). *a.* Habit, X½. *b.* Leaf, X10. *c.* Fruit, X7½.

5 (–10) cm long; leaves linear to linear-oblong, 1–3 cm long, 0.8–2.5 mm wide, each margin with 3–12 coarse teeth, the sheath distended, entire, or 1- to 2-toothed on each margin; staminate flower 3–4 mm long, rupturing from spathe; pistillate flower 3.0–3.5 mm long; achene ellipsoid, 4.0–7.5 mm long (excluding the beak), 1.5–3.0 mm broad, the surface pitted, the slender recurved beak 1.0–1.5 mm long.

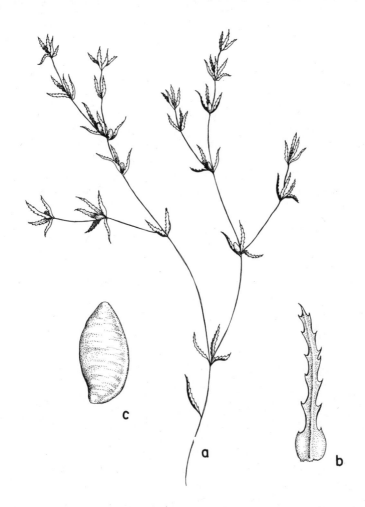

51. Najas marina (Naiad). *a.* Habit, X¼. *b.* Leaf, X3. *c.* Achene, X4½.

COMMON NAME: Naiad.

HABITAT: In three feet of water; pH 8.4 (in Illinois).

RANGE: Widely scattered throughout the United States; Mexico; Central America; Europe; Asia.

ILLINOIS DISTRIBUTION: Known only from Lake County (Druce Lake, September 9, 1964, *G. Tichacek s.n.;* 2½ miles NE of Grayslake, October 5, 1964, *G. S. Winter-ringer 22499* and *22503;* Wooster Lake, August 4, 1964, *B. Dolbeare 311* and *312*).

The coarsely toothed leaves and larger fruits of this dioecious species readily distinguish it from the other members of this genus in Illinois. It is the only member of subgenus *Najas* in the Illinois flora.

Order Arales

Flowers perfect or unisexual, usually associated with a fleshy spadix subtended by a foliaceous spathe; perianth parts 0, 4, or 6, none of them petal-like; stamens 1, 2, 4, or 6; ovary 1, superior or partially embedded in the spadix.

The two families comprising this order have no obvious resemblance to each other. The Lemnaceae is thought to be a highly reduced family of araceous ancestry.

KEY TO THE FAMILIES OF Araceae IN ILLINOIS

1. Stems and leaves well-developed; spadix present; usually robust perennials_____Araceae, page 124
1. Plant body thalloid; spadix absent; free-floating minute aquatics__ _____Lemnaceae, page 138

ARACEÆ–ARUM FAMILY

Herbaceous perennials from corms, rhizomes, or thick, fleshy roots; leaves sometimes net-veined, simple or compound; flowers unisexual or perfect, attached to a fleshy spadix; perianth parts 0, 4, or 6; stamens 2–6; staminodia sometimes present; ovary 1- to 3-celled, with 1–several ovules per cell; fruit dry, or fleshy and berry-like, 1- to 3-seeded.

Many genera of this primarily tropical family are cultivated as houseplants. Included here are *Philodendron, Monstera,* the cut-leaved philodendron, *Aglaonema,* the Chinese evergreen, *Pothos,* and *Dieffenbachia,* the dumb cane.

KEY TO THE GENERA OF Araceae IN ILLINOIS

1. True spathe absent; ovary 2- to 3-celled; fruit dry; leaves linear, the midvein not central_____1. *Acorus*
1. Spathe present; ovary 1-celled; fruit fleshy; leaves broad, the midvein central.
    2. Flowers perfect; spadix nearly globoid; perianth parts 4; fruit 8–12 cm in diameter; leaves simple, ovate; plants with rhizomes _____2. *Symplocarpus*
    2. Flowers unisexual; spadix elongate; perianth absent; fruit to 6 cm long, narrower; leaves simple, arrowhead-shaped, or compound; plants with corms or fleshy roots.
        3. Staminodia present; berries brownish or greenish; leaves simple, arrowhead-shaped.

4. Plants rhizomatous; spathe tubular only at base; flowers confined to the lower half of the spadix_____3. *Arum*

4. Plants with fleshy roots; spathe tubular at both ends, opening at the middle; flowers covering all or most of the spadix_____4. *Peltandra*

3. Staminodia absent; berries red; leaves compound_____
_____5. *Arisaema*

1. *Acorus* L. – Sweet Flag

Plants rhizomatous; leaves linear, with an excentric midvein; true spathe none; spadix elongate, completely covered with perfect flowers; perianth parts 6; stamens 6; ovary 2- to 3-celled; fruit dry, 1- to 3-seeded.

Only the following species occurs in Illinois.

1. Acorus calamus L. Sp. Pl. 324. 1753. *Fig. 52.*

Rhizome stout, creeping, aromatic; leaves numerous, basal, linear, to 1 m long, to 2 cm broad; scape flattened, leaf-like, to 1 m long, extended to 60 cm beyond the spadix as a modified, open spathe; spadix to 10 cm long, 0.8–2.0 cm thick; flowers small, brownish-yellow, the perianth segments concave.

COMMON NAME: Sweet Flag.

HABITAT: Marshes and other low areas.

RANGE: Nova Scotia to Oregon, east and south to Colorado, Texas, and Florida; Europe; Asia.

ILLINOIS DISTRIBUTION: Occasional; scattered in all parts of the state.

Flowering extends over a long period from late May to mid-August. The aromatic and sweet-tasting rhizome has had reputed medicinal properties.

This genus is unusual among the Araceae in Illinois by the absence of a true spathe and by the presence of a 6-parted perianth and 6 stamens.

2. *Symplocarpus* SALISB. – Skunk Cabbage

Plants rhizomatous; leaves ovate, net-veined; spathe fleshy, nearly completely enclosing the globoid spadix; flowers perfect, completely covering the spadix; perianth parts 4; stamens 4; ovary 1-celled, buried in the spadix; fruit fleshy, the seeds embedded.

Only the following species comprises the genus.

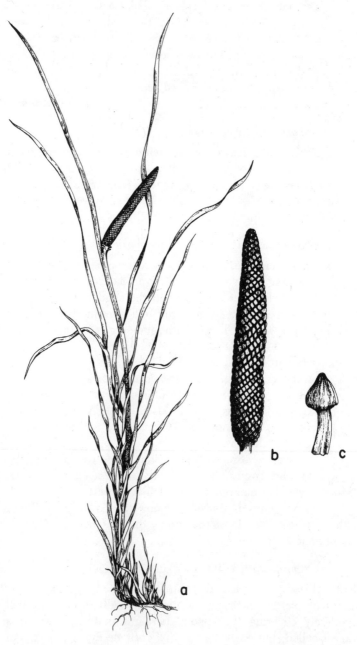

52. *Acorus calamus* (Sweet Flag). *a.* Habit, X⅛. *b.* Spadix, X⅜. *c.* Fruit, X2½.

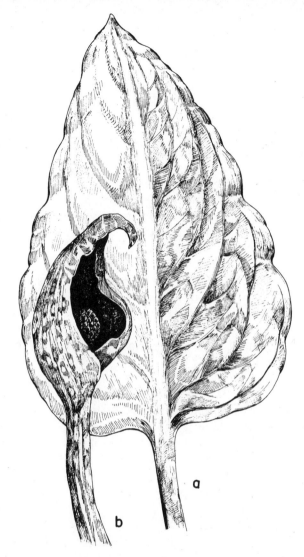

53. *Symplocarpus foetidus* (Skunk Cabbage). *a.* Leaf, X¼. *b.* Spathe and spadix, X¼.

1. Symplocarpus foetidus (L.) Nutt. Gen. 1:105. 1818. *Fig. 53.*

Dracontium foetidum L. Sp. Pl. 967. 1753.

Spathyema foetida (L.) Raf. Med. Repos. II. 5:352. 1808.

Rhizome stout, more or less erect; leaves appearing after the flowers, numerous, basal, ovate, obtuse at apex, cordate, net-veined, to 45 (–50) cm long, to 35 (–40) cm broad; petiole rather stout, much shorter than the blades; spathe ovoid, curved above, the margins inrolled, to 15 cm long, green with purple spots and stripes; spadix globoid; perianth parts becoming fleshy, persistent on the fruit and strongly odorous; fruit 8–12 cm in diameter, spongy, odorous, the seeds embedded, 8–10 mm in diameter; $2n = 30$ (Ito, 1942).

COMMON NAME: Skunk Cabbage.

HABITAT: Swamps and other low areas.

RANGE: Quebec to Manitoba, south and east to Tennessee and Georgia; Asia.

ILLINOIS DISTRIBUTION: Occasional; restricted to the northern three-fifths of the state.

The fleshy fruit has a putrid odor resembling that of a skunk. The flowers appear in February and March, before the leaves. In the northern third of Illinois, skunk cabbage is rather frequent in low wet areas. During summer, there is a gigantic expansion in the size of some of the leaves.

3. *Arum* L. – Arum

Perennials from rhizomes; leaves sagittate, pinnately veined; spathe elongated, tubular at base, often showy; spadix floriferous in the lower half, sterile above and prolonged as a cylindrical appendage; flowers unisexual, ebracteate, the lowermost fertile; perianth 0; fruit a berry.

Only the following species occurs in Illinois.

1. Arum italicum Mill. Gard. Dict. ed. 8:2. *Fig. 54.*

Leaves basal, several, more or less acute at apex, sagittate-hastate at base, to 45 cm long at flowering time, often becoming much larger later, nearly as broad; petiole stout; spathe to 30 cm long, tubular at base, whitish-green sparsely dotted with purple; spadix elongated, the upper part cylindrical and sterile, yellow; staminate flowers usually above the pistillate ones, usually with pistillodia; pistillate flowers with staminodia; berry greenish or brownish.

54. Arum italicum (Arum). X¼.

COMMON NAME: Arum.

RANGE: Native of western Europe and North Africa; adventive in Illinois.

ILLINOIS DISTRIBUTION: Escaped from cultivation in a woodland on the campus of Southern Illinois University (Jackson County) where it has persisted for several years. The leaves of this species appear early in spring and are about two-thirds grown when the flowers are borne in early May. The locality cited above has been in existence since at least 1958. Mohlenbrock and Ozment (1967) erroneously reported this species as *Arum exoticum* L.

4. *Peltandra* RAF. – Arrow Arum

Perennial from thick, fibrous roots; leaves sagittate, pinnately veined; spathe elongated, usually convolute below and above, open near the middle, enclosing the elongated spadix; perianth none; stamens 4–5, embedded in sterile ovary tissue; staminodia 4–5, scale-like; ovary 1-celled; fruit a cluster of berries, 1- to 3-seeded.

Only the following species occurs in Illinois.

1. **Peltandra virginica** (L.) Kunth, Enum. 3:43. 1841. *Fig. 55.*

Arum virginicum L. Sp. Pl. 966. 1753.

Peltandra undulata Raf. Journ. Phys. 89:103. 1819.

Leaves basal, several, acute at apex, sagittate to hastate at base, variable in shape, to 30 cm long at flowering time, enlarging later, to 18 cm broad; petiole rather stout, up to 45 cm long; spathe 10–20 cm long, green with pale margin, convolute at both ends, open at the middle exposing staminate flowers; spadix elongated, almost entirely covered by the flowers, the staminate flowers apical, soon falling away with the upper part of the spadix, the pistillate flowers basal; berries greenish to brown, to 12 mm long, nearly as broad.

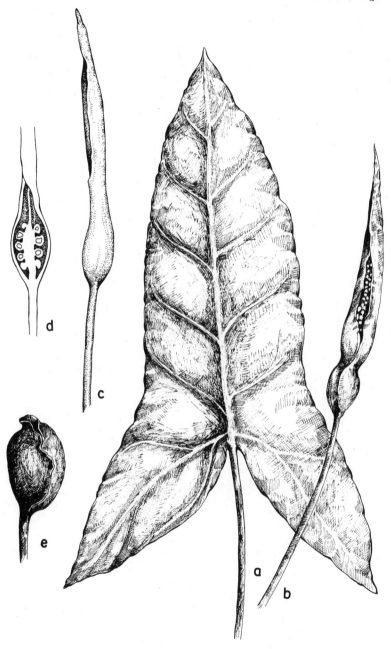

55. *Peltandra virginica* (Arrow Arum). *a.* Leaf, X¼. *b.* Spadix and spathe, X¼. *c.* Spathe, X¼. *d.* Lower part of spadix (some of spathe cut away), X⅓. *e.* Fruit, X1½.

COMMON NAME: Arrow Arum.

HABITAT: Swamps, shallow water, wet ditches.

RANGE: Maine to Michigan, south to Texas and Florida.

ILLINOIS DISTRIBUTION: Occasional; absent in the northern one-fifth of the state.

Leaf shape is highly variable. Many of these variations have been recognized as forms by Blake (1912), but intergradation is so complex that there is little value in recognizing these segregates. The flowers are borne in May and June.

Peltandra is the only Illinois genus of Araceae with fibrous roots. Like *Arisaema,* there are no remnants of the perianth remaining.

5. *Arisaema* MART. – Jack-in-the-Pulpit

Perennial from corms; leaves compound, sheathing at the base; spathe elongated, convolute below, arched above; spadix elongated; perianth none; stamens 2–5, nearly sessile; ovary 1-celled; fruit a cluster of berries, 1- to 3-seeded.

KEY TO THE SPECIES OF Arisaema IN ILLINOIS

1. Leaflets (3–) 5–13; spathe convolute below and above, open near the middle; spadix slender-tapering, exserted beyond the spathe _____1. *A. dracontium*
1. Leaflets 3; spathe convolute below, open and arching above; spadix club-shaped, obtuse, covered by the arching spathe_____ _____2. *A. triphyllum*

1. **Arisaema dracontium** (L.) Schott in Schott & Endlicher, Melet. 1:17. 1832. *Fig. 56.*

Arum dracontium L. Sp. Pl. 964. 1753.

Maricauda dracontium (L.) Small, Fl. S.E. U. S. 227. 1903.

Leaf usually solitary, with (3–) 5–13 leaflets, the petiole sometimes nearly 1 m long at maturity; leaflets elliptic to oblong-lanceolate, acuminate at apex, tapering to the base, the terminal 10–18 cm long, the others somewhat smaller, pinnately-veined; peduncle arising from the leaf sheath, shorter than the petiole; spathe slender, convolute below and above, to 6 cm long, acuminate, green; spadix elongated, long-tapering, exserted 5–10 cm beyond the spathe; fruiting head narrowly conical, with

56. Arisaema dracontium (Green Dragon). *a.* Habit, X¼. *b.* Fruit, X¼.

several globoid orange-red berries 2.5–5.0 mm in diameter; 2n = 56 (Bowden, 1940).

COMMON NAME: Green Dragon.

HABITAT: Rich woodlands.

RANGE: Quebec to Minnesota, south to Texas and Florida.

ILLINOIS DISTRIBUTION: Common; in every county. The flowers appear in May and June, with fruit maturing in August and September. Variation occurs in size of the plant and number and shape of the leaflets. A remarkable colony of specimens with three leaflets has been found near Gorham in Jackson County.

Rapid expansion of the leaves usually follows flowering of the green dragon.

2. Arisaema triphyllum (L.) Schott in Schott & Endlicher, Melet. 1:17. 1832.

Leaves usually 2, with 3 leaflets, the petiole sometimes becoming over 1 m tall at maturity; leaflets broadly elliptic to ovate, acute to acuminate at the apex, more or less tapering to the usually asymmetrical base, the terminal larger than the lateral, pinnately-veined; peduncle arising from one of the leaf sheaths, shorter than the petiole; spathe convolute below, open and arching above, green to purple, one-colored or striped, short-acuminate; spadix club-shaped, obtuse, yellowish, covered by the arching spathe, naked at the apex, with pistillate flowers basal and staminate flowers above them; fruiting head ovoid or globoid, with several bright red berries 8–10 mm in diameter.

The corms of this species are exceedingly sharp to the taste.

The taxonomy of *A. triphyllum* is not agreed upon by taxonomists. Gleason (1952) recognizes three varieties for the eastern United States, while Fernald (1950), following Huttleston (1949), recognizes three species and four forms.

The Illinois material seems to fall into two varieties.

a. Lateral leaflets strongly asymmetrical; expanded portion of spathe 4–7 cm wide, green or striped with purple, rarely solidly purple; fruiting head 3–6 cm long_____2a. *A. triphyllum* var. *triphyllum*

a. Lateral leaflets scarcely asymmetrical; expanded portion of spathe to 3 cm wide, solidly purple; fruiting head to 2 cm long_____ _____2b. *A. triphyllum* var. *pusillum*

2a. Arisaema triphyllum (L.) Schott var. **triphyllum** *Fig. 57.*

Arum triphyllum L. Sp. Pl. 965. 1753.

Arum atrorubens Ait. Hort. Kew. 3:315. 1785.

Arum triphyllum a zebrinum Sims, Curtis' Bot. Mag. t. 950. 1806.

Arisaema atrorubens (Ait.) Blume, Rumphia 1:97. 1835.

Arisaema atrorubens var. *viride* Engler in DC. Monogr. 1:536. 1879.

Lateral leaflets strongly asymmetrical; expanded portion of spathe 4–7 cm wide, green or striped with purple, rarely solidly purple; fruiting head 3–6 cm long: 2n = 28, 56 (Huttleston, 1949).

COMMON NAME: Jack-in-the-Pulpit; Indian Turnip.

HABITAT: Rich woods.

RANGE: New Brunswick to Manitoba, south to Kansas and Florida.

ILLINOIS DISTRIBUTION: In every Illinois county. Jack-in-the-pulpit is a favorite among wild flower enthusiasts. The common variety is extremely variable, particularly with reference to color of the spathe. Solid purple or solid green spathes are found occasionally, although the striped spathe is encountered most frequently.

2b. Arisaema triphyllum (L.) Schott var. **pusillum** Peck, Rep. N. Y. State Mus. 51:297. 1898. *Fig. 58.*

Arisaema pusillum (Peck) Nash. in Britt. Man. Fl. N. States 229. 1901.

Lateral leaflets scarcely asymmetrical; expanded portion of spathe to 3 cm wide, solidly purple; fruiting head to 2 cm long.

COMMON NAME: Small Jack-in-the-Pulpit.

HABITAT: Rich woodlands.

RANGE: New York to Illinois, south to Georgia.

ILLINOIS DISTRIBUTION: Known only from Alexander County (2 miles west of Tamms, May 8, 1951, *H. E. Ahles 3848*).

The Small Jack-in-the-Pulpit is smaller in all respects than the common variety. It may not always show clear-cut differences.

57. *Arisaema triphyllum* var. *triphyllum* (Jack-in-the-Pulpit). *a.* Habit, X¼. *b.* Fruit, X¼.

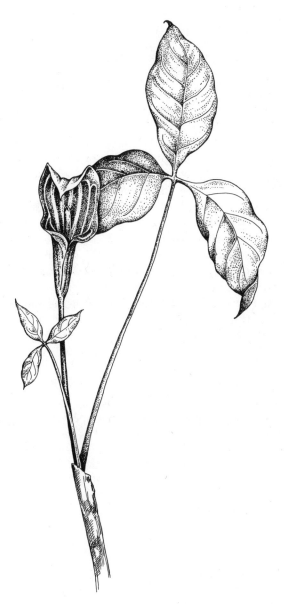

58. *Arisaema triphyllum* var. *pusillum* (Small Jack-in-the-Pulpit). X¼.

LEMNACEÆ–DUCKWEED FAMILY

Free-floating thalloid plants with o-several ventral rootlets originating at the node; fronds nerved or nerveless; vegetative buds with flowers produced at the node in lateral, basal, or dorsal reproductive pouches; plants monoecious; flowers, in some species, surrounded by a spathe; staminate flowers 1–2 (–3), the anthers 1- to 2-celled; pistillate flowers 1, the ovary 1-celled, with 1 to several ovules; fruit a utricle, frequently ribbed; seed smooth or ribbed.

The aquatic family Lemnaceae has presented serious taxonomic difficulties for many years. The family has been monographed several times (Schleiden, 1839; Hegelmaier, 1868; Daubs, 1965), and a significant revision of the North American taxa excluding Mexico was completed by Thompson in 1898. Despite the existance of these major contributions, several closely related species remain recognized only questionably.

Flowering and fruiting specimens are extremely rare in collections and, when they do occur, are so reduced as to offer few characters of classificatory value. Thus, the delimitation of both the genera and the species rests heavily on vegetative features. It is partly because of this, since vegetative characters are variable, that taxonomic difficulties occur in the family.

The term frond is employed when speaking of the thalloid portion of the plant not including rootlets, flowers, and fruits. The term plant is more appropriately used when referring to the latter organs and the frond as a unit. The upper surface of the frond is referred to as the dorsal surface, while the lower surface is designated the ventral surface.

For a detailed account of the Illinois species, see Weik and Mohlenbrock (1968).

KEY TO THE GENERA OF Lemnaceae IN ILLINOIS

1. Rootlets 1-several per plant; reproductive pouches lateral, 2 per frond.
    2. Rootlets (1–) 2 or more per plant_____1. *Spirodela*
    2. Rootlet 1 per plant_____2. *Lemna*
1. Rootlets none; reproductive pouch basal, 1 per plant.
    3. Frond thin, linear_____3. *Wolffiella*
    3. Frond thick, globose or ellipsoid_____4. *Wolffia*

1. *Spirodela* SCHLEIDEN

Fronds flattened, orbicular to reniform, usually asymmetrical,

3- to 8-nerved; rootlets (1–) 2–10, arising in a fascicle beneath the node; reproduction from two lateral pouches on either side of the node; flowers and fruit infrequently produced; ovules 2.

KEY TO THE SPECIES OF Spirodela IN ILLINOIS

1. Fronds orbicular, 5- to 8-nerved; rootlets 2–10_____1. S. polyrhiza
1. Fronds obovate to reniform, obscurely 3- (5-) nerved; rootlets (1–) 2–5_____2. S. oligorhiza

1. Spirodela polyrhiza (L.) Schleiden, Linnaea 13:392. 1839.
   Fig. 59 (a–c).

Lemna polyrhiza L. Sp. Pl. 970. 1753.

Lemna major Griff. Notul. 3:216. 1851.

Fronds orbicular to obovate (immature fronds orbicular), weakly to strongly asymmetrical, apiculate, 2–7 mm long, 2–6 mm wide, weakly inflated, cavernous throughout, prominently 5- to 8-nerved, the nerves originating palmately at the node, lateral nerves strongly incurved toward the central nerve after departure from the node; upper surface flattened, often olive-green, guard cells abundant, with a row of papules frequently present along the central nerve and occasionally along lateral nerves; lower surface flattened or weakly convex, pale green or occasionally reddish-purple; rootlets 2–10, arising in a fascicle directly beneath the node on the lower surface, up to 15 mm long, rootcaps acute; plants solitary or commonly remaining attached in groups of 2–5; vegetatively produced plants and infrequently observed flowers arising from two lateral, reproductive pouches; fruit a somewhat compressed and rounded utricle with winged margins; seed smooth or only slightly ribbed.

COMMON NAME: Duckweed.

HABITAT: Standing water.

RANGE: Throughout North America; tropical America; Europe; Asia.

ILLINOIS DISTRIBUTION: Throughout the state, probably in every county.

This species is variable in both shape and thickness, but despite this variability is easily separated from the other species upon critical examination of nerve and rootlet numbers. The frond of S. polyrhiza is also generally larger than that of S. oligorhiza.

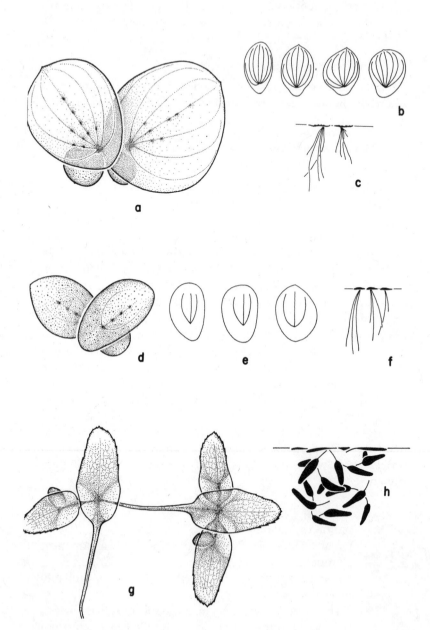

The nodal region on the upper surface of the frond is often red-pigmented and has been referred to by various authors as the "eyespot." According to Hicks (1937), this condition is most prevalent in fronds growing in the sun.

2. Spirodela oligorhiza (Kurtz) Hegelm. Die Lemnaceen 147. 1868. *Fig. 59* (d–f).

Lemna oligorhiza Kurtz, Journ. Linn. Soc. London 9:267. 1867.

Fronds ovate to reniform, asymmetrical, subacute to obtuse at the apex, 2.2–6.0 mm long, 2.0–2.6 mm wide, inflated, cavernous throughout, obscurely 3- (5-) nerved; upper surface convex, sparsely punctate with pigment cells, guard cells abundant; lower surface weakly convex, somewhat darker than the upper; rootlets (1–) 2–5, arising in a darkened fascicle directly beneath the node; rootcaps obtuse; plants solitary or commonly remaining attached in groups of 2–3; vegetatively produced plants and infrequently observed flowers arising from two lateral, reproductive pouches; fruit slightly winged, ovoid; seeds longitudinally ribbed.

COMMON NAME: Duckweed.

HABITAT: Standing water.

RANGE: Native in the Orient; supposedly sparingly introduced in the United States.

ILLINOIS DISTRIBUTION: Known from only Alexander and Union counties in the extreme southern tip of the state.

Spirodela oligorhiza is easily distinguished from *S. polyrhiza*, but the inexperienced student might mistake it for a member of the genus *Lemna* since at times it may possess only a single rootlet.

The biochemical data of McClure and Alston (1966) strongly suggest a close relationship between *S. oligorhiza* and *L. minor*. This relationship was recognized by Hegelmaier, also (1896).

59. *Spirodela polyrhiza* (Duckweed). *a.* Habit, dorsal aspect, X2½. *b.* Morphological variation among mature fronds, X1. *c.* Floating habit, lateral aspect, X½. *Spirodela oligorhiza* (Duckweed). *d.* Habit, dorsal aspect, X2½. *e.* Morphological variation among mature fronds, X1. *f.* Floating habit, lateral aspect, X½. *Lemna trisulca* (Duckweed). *g.* Habit, dorsal aspect, X2½. *h.* Floating habit, X½.

Future evidence may necessitate a more realistic positioning of this species in the genus *Lemna*.

2. Lemna L. – Duckweed

Fronds flattened to strongly convex on the ventral surface, orbicular to spatulate, symmetrical to asymmetrical, occasionally long-stipitate, 1- to 3- (5-) nerved, the nerves frequently obscure; rootlet 1, the rootcap acute to obtuse; reproduction from two lateral pouches on either side of the node; flowers and fruit sparingly produced; ovules 1–6.

KEY TO THE SPECIES OF Lemna IN ILLINOIS

1. Fronds spatulate with long, persistent stipes, often submerged in compact masses; rootlet frequently absent_____1. *L. trisulca*
1. Fronds orbicular to elliptic, floating; rootlet present.
 2. Fronds 3- to 5-nerved.
  3. Root sheaths cylindrical.
   4. Lower surface of frond flattened or only weakly convex, apex rounded to acute, frond usually 3-nerved_____ _____2. *L. minor*
   4. Lower surface of frond moderately to strongly convex, apex broadly rounded or often obtuse, frond 3- to 5-nerved _____3. *L. gibba*
  3. Root sheaths winged.
   5. Fronds weakly to strongly asymmetrical, thick, the nerves often inconspicuous_____4. *L. perpusilla*
   5. Fronds symmetrical or nearly so, membranous, the 3 nerves prominent_____5. *L. trinervis*
 2. Fronds 1-nerved or obscurely nerved.
  6. Fronds 1-nerved, the lower surface green.
   7. Fronds ovate-elliptic to elliptic, weakly to moderately asymmetrical, often floating in compact masses_____ _____6. *L. valdiviana*
   7. Fronds obovate to orbicular, symmetrical, usually solitary _____7. *L. minima*
  6. Fronds obscurely nerved, the lower surface frequently reddish-purple_____8. *L. obscura*

1. **Lemna trisulca** L. Sp. Pl. 970. 1753. *Fig.* 59 (g–h).

Fronds elliptic to oblanceolate, commonly falcate with long-tapering stipes, 4–10 mm long, 1.5–3.0 mm wide, membranous, obscurely 3-nerved, the apical margin serrulate; upper surface flattened, dull green, guard cells absent; lower surface flattened; rootlets normally lacking, if present, the length highly variable, at times weakly coiled, the sheath winged, the rootcap long, obtuse; plants attached to one another at the nodes by elongated stipes, rarely solitary, generally submerged; fruit symmetrical; seeds longitudinally ribbed with numerous cross-striae.

COMMON NAME: Duckweed.
HABITAT: Standing water.
RANGE: North America; Europe; Asia; Africa.
ILLINOIS DISTRIBUTION: Local throughout the state.
This species is often referred to as star duckweed or submerged duckweed because attached plants frequently produce stellate colonies which are normally submerged. It is the most easily recognized *Lemna* in Illinois.

2. **Lemna minor** L. Sp. Pl. 970. 1753. *Fig. 60* (a–e).

Fronds obovate to elliptical (immature fronds orbicular), symmetrical or weakly asymmetrical, 2–5 mm long, 1.5–3.5 mm wide, flattened to weakly inflated, cavernous throughout, 3-nerved, the nerves occasionally obscure; upper surface usually slightly convex, with a thick, glistening cuticle, guard cells abundant, a median row of papules commonly present; lower surface flattened or weakly to moderately convex, pale green or occasionally reddish-purple; rootlet one, arising obliquely beneath the node and lying in a narrow furrow, its length highly variable, the sheath frail, cylindrical, the rootcap obtuse; plants solitary or commonly remaining attached in groups of 2–4; fruit symmetrical, broadly ovoid; seeds longitudinally ribbed.

COMMON NAME: Duckweed.
HABITAT: Standing water.
RANGE: North America; Europe; Asia; Africa; Australia.
ILLINOIS DISTRIBUTION: Throughout the state.
This species is apparently quite variable in both size and coloration. Large specimens have been observed up to 5 mm in length. These types are usually somewhat flattened and membranous. Other specimens have been observed with an obvious reddish-purple

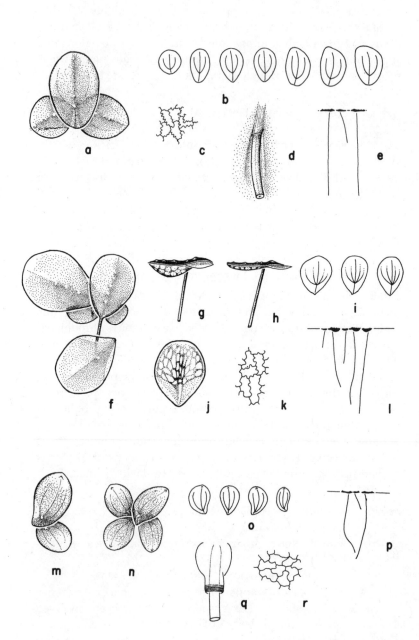

pigmentation on the lower surface of the frond. This condition appears to be associated with the smaller plants which also usually have conspicuously inflated fronds. The latter specimens are often quite difficult to separate from *L. obscura;* however, careful examination normally reveals three obscure nerves.

3. **Lemma gibba** L. Sp. Pl. 970. 1753. *Fig. 60* (f–l).

Fronds orbicular to cuneate, symmetrical to weakly asymmetrical, 2–6 mm long, 1–4 mm wide, commonly inflated and cavernous throughout especially on the lower surface, obscurely 1- to 3- (5-) nerved; upper surface flat to weakly convex, the guard cells abundant, commonly with a median row of papules and occasionally keeled, dark green or sometimes purple; lower surface flattened to strongly gibbous; rootlet one, the sheath cylindrical, prominent, the rootcap obtuse; plants solitary or commonly remaining attached in groups of 2–4 by short to somewhat elongated stipes; fruit winged, depressed-globoid; seeds 2–4, compressed, asymmetrical, longitudinally ribbed.

COMMON NAME: Duckweed.
HABITAT: Standing water.
RANGE: North America; South America; Europe; Asia; Africa; Australia.
ILLINOIS DISTRIBUTION: Known from three counties in the northern half of the state (Knox: *McClure 46;* Mason: *Burrill s.n.;* Will: *Daubs 877*).
The first collection of this duckweed in Illinois was apparently made in Mason County by T. W. Burrill in 1894. Although Burrill's specimens are dried, their

60. *Lemna minor* (Duckweed). *a.* Habit, dorsal aspect, X2½. *b.* Morphological variation among mature fronds, X1¾. *c.* Epidermal cell wall pattern, lower surface, X100. *d.* Origin of single rootlet showing characteristic furrow formed on lower surface, X6¼. *e.* Floating habit, lateral aspect, X½. *Lemna gibba* (Duckweed). *f.* Habit, dorsal aspect, X2½. *g.* Lateral view of a frond showing strong development of gibbous condition on lower surface, X2½. *h.* Lateral view of a frond showing weak development of gibbous condition on the lower surface, X2½. *i.* Morphological variation among mature fronds, X1¾. *j.* Lower surface of a frond showing development of air spaces in strongly gibbous condition, X2½. *k.* Epidermal cell wall pattern, lower surface, X100. *l.* Floating habit, lateral aspect, X½. *Lemna perpusilla* (Duckweed). *m. & n.* Habit, dorsal aspects, X2½. *o.* Morphological variation among mature fronds, X1¾. *p.* Floating habit, X½. *q.* Winged sheath surrounding rootlet at point or origin on lower surface, X12½. *r.* Epidermal cell wall pattern, lower surface, X100.

strong resemblance to living plants of *L. gibba* is unquestionable. The specimens of McClure and Daubs leave no doubt concerning the authenticity of its occurrence in Illinois.

Upon drying, specimens of *L. gibba* may collapse enough to resemble superficially *L. minor* or *S. polyrhiza*.

4. **Lemna perpusilla** Torr. Fl. N. Y. 2:245. 1843. *Fig. 60* (m–r).

Fronds ovate to obovate, variable, asymmetrical, 1.5–3.0 mm long, 1.0–1.5 mm wide, inflated, obscurely 3-nerved or apparently nerveless, cavernous in the central region only; upper surface weakly convex, the guard cells abundant, with prominent nodal and apical papules and occasional smaller papules on median line between; lower surface flattened or weakly convex; rootlet one, with a winged sheath, the root cap acute; plants solitary or commonly remaining attached in groups of 2–3; fruit asymmetrical, ovoid to oblongoid with a prominent oblique style directed toward the apex of the frond; seeds longitudinally ribbed and with numerous cross-striae.

COMMON NAME: Duckweed.

HABITAT: Standing water.

RANGE: Massachusetts to South Dakota, south to Kansas and Florida; California; South America.

ILLINOIS DISTRIBUTION: Scattered throughout the state, but not commonly encountered.

The characteristic winged root sheath and asymmetrical frond beset with papules help to distinguish this duckweed from all others in Illinois. The winged root sheath can best be seen with the aid of a compound microscope.

5. **Lemna trinervis** (Austin) Small, Fl. S.E. U. S. 230. 1903. *Fig. 61* (a–e).

Lemna perpusilla var. *trinervis* Austin in Gray, Man. Bot. 479. 1867.

Fronds obovate to ovate-elliptic, symmetrical, 2.5–4.0 mm long, 1.5–2.0 mm wide, relatively thin and membranous, prominently 3-nerved, weakly cavernous in the central region only; upper surface flattened, the guard cells abundant, an apical papule sometimes present; lower surface flattened; rootlet one, with a winged sheath, the rootcap acute; plants solitary or commonly remaining attached in groups of 2–3; fruit asymmetrical, broadly

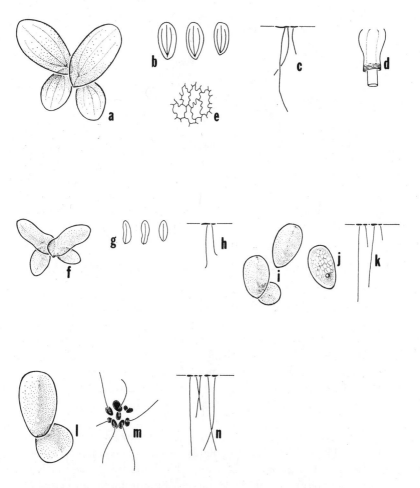

61. *Lemna trinervis* (Duckweed). *a.* Habit, dorsal aspect, X2½. *b.* Morphological variation among mature fronds, X1¼. *c.* Floating habit, lateral aspect, X½. *d.* Winged sheath surrounding rootlet at point of origin on lower surface, X12½. *e.* Epidermal cell wall pattern, lower surface, X100. *Lemna valdiviana* (Duckweed). *f.* Habit, dorsal aspect, X2½. *g.* Morphological variation among mature fronds, X1¾. *h.* Floating habit, lateral aspect, X½. *Lemna minima* (Duckweed). *i.* Habit, dorsal aspect, X2½. *j.* Lower surface of a frond showing inflated central region, X2½. *k.* Floating habit, lateral aspect, X½. *Lemna obscura* (Duckweed). *l.* Habit, dorsal aspect, X2½. *m.* Lower surface of a frond showing dark pigmentation, X½. *n.* Floating habit, lateral aspect, X½.

ovoid to oblongoid with a prominent oblique style directed toward the apex of the frond; seeds longitudinally ribbed with numerous cross-striae.

COMMON NAME: Duckweed.
HABITAT: Standing water.
RANGE: North America; South America.
ILLINOIS DISTRIBUTION: Throughout the state, but not often encountered.
The morphological similarity between *L. perpusilla* and *L. trinervis* is obvious. Several important characteristics are shared by both taxa. The frond of *L. trinervis*, however, tends to be thin and membranous and its three nerves are very prominent. Its symmetry is normally regular, although at times it may exhibit a slight asymmetry more characteristic of *L. perpusilla*.

6. **Lemna valdiviana** Phil. Linnaea 33:239. 1864. *Fig. 61* (f–h).

Lemna cyclostasa Ell. ex Thomp. Rep. Mo. Bot. Gard. 9:35. 1898.

Fronds oblong to oblong-elliptic, asymmetrical and often distinctly falcate, 1–5 mm long, 0.5–2.0 mm wide, obscurely 1-nerved or nerveless, cavernous throughout; upper surface convex, usually pale green and glossy, with guard cells sparingly produced, a median row of minute papules occasionally present; lower surface flattened or weakly convex, pale green; rootlet one, the sheath long and cylindrical, the rootcap usually long and weakly reflexed, acute; plants usually remaining attached in groups of 2–4 or more, frequently forming dense masses; fruit asymmetrical, oblongoid; seeds longitudinally ribbed and with numerous cross-striae.

COMMON NAME: Duckweed.
HABITAT: Standing water.
RANGE: United States; South America.
ILLINOIS DISTRIBUTION: In the southern one-third of Illinois; also Lake County.
Lemna valdiviana may be distinguished from other members of the genus by its narrow, often falcate fronds and generally smaller size, although very small plants may occasionally resemble *L. minima*. McClure and Alston (1966), however, found these two taxa to be quite distinct in their flavonoid chemistry.

7. **Lemna minima** Phil. Linnaea 33:239. 1864. *Fig. 61* (i–k).

Lemna valdiviana var. *minima* (Phil.) Hegelm. Lemnac. 138. 1868.

Fronds obovate to elliptical, symmetrical to weakly asymmetrical, 1.5–1.7 mm long, 1.0–1.2 mm wide, weakly inflated, obscurely 1-nerved or apparently nerveless, cavernous only in the central region, thin-margined; upper surface convex, pale green, glossy, guard cells present, with a median row of papules commonly present; lower surface flattened or weakly convex, pale green or yellowish; rootlet one, the sheath cylindrical, thin, the rootcap acute or obtuse; plants frequently solitary or remaining attached in groups of 2 (–4); fruit elongated, symmetrical; seeds longitudinally ribbed and with cross-striae.

COMMON NAME: Duckweed.

HABITAT: Standing water.

RANGE: Minnesota to California, south to Texas and Florida; Mexico; Central America; South America.

ILLINOIS DISTRIBUTION: Known only from three widely scattered counties (Carroll: *Jones 17292;* Madison: *Evers 31168;* Will: *Hill 183*).

This species probably presents the greatest difficulty for students of Illinois duckweeds. Its small size and poor condition upon drying contribute much to the problem of proper identification. It is often erroneously identified as *L. minor* or *L. perpusilla.* Jones (1963) circumvents this problem by combining this species with *L. minor* in the *Flora of Illinois.* The flavonoid chemistry of these two species is so distinct, however, that evidence is strongly in favor of retaining *L. minima* as a legitimate species.

8. **Lemna obscura** (Austin) Daubs, Ill. Biol. Mon. 34:20. 1965. *Fig. 61* (l, m, & n).

Lemna minor var. *obscura* Austin in Gray, Man. Bot. 479. 1867.

Fronds obovate to elliptical, 1.5–2.0 mm long, 1.0–1.5 mm wide, obscurely nerved, moderately inflated and cavernous throughout, thin margined; upper surface flattened to weakly convex, pale green or sometimes slightly reddish, guard cells present, occasionally with a median row of small papules; lower surface moderately to strongly convex, reddish-purple; rootlet one, the sheath cylindrical, reduced, the plants solitary or remaining attached in groups of 2(-4); rootcap obtuse; fruit obovoid, turbinate; seeds longitudinally ribbed, usually red-pigmented.

COMMON NAME: Duckweed.

HABITAT: Standing water.

RANGE: United States; Bahamas; Mexico.

ILLINOIS DISTRIBUTION: Primarily in the central third of the state.

This species may not be distinct from *L. minor*, although the data of McClure and Alston (1966) support its recognition. In collections examined from Mason (*R. T. Rexroat 750* and *2222*) and Menard (*R. T. Rexroat 5615*) counties, typical *L. minor* was found mixed with *L. obscura*.

Dried fronds of *L. obscura* are usually pale olive-green and remain inflated, the thin margins frequently becoming involute. A small apical papule may also become prominent.

3. *Wolffiella* HEGELMAIER

Fronds thin, linear, nerveless; rootlets absent; vegetative reproduction from a single, basal, reproductive pouch, the flowers and fruit rare, produced on the upper surface, ovule one; young plants remaining attached to form stellate colonies.

Only the following species occurs in Illinois.

1. **Wolffiella floridana** (J. D. Smith) Thompson, Rep. Mo. Bot. Gard. 9:37. 1898. *Fig. 62* (a–c).

Wolffiella gladiata var. *floridana* J. D. Smith, Bull. Torrey Club 7:64. 1880.

Fronds thin, linear, attenuate, with an acute or obtuse apex, often falcate and curving downward when floating, 3–12 mm long, 0.5–1.0 mm wide, cavernous throughout, nerveless, guard cells rarely present; plants mostly submerged, solitary or commonly remaining attached in 2–4 or more membered groups, often forming stellate colonies where many members are coherent; new plants arising from a ring, the basal node situated in a wedge-shaped, often spreading, membranous, basal reproductive pouch; flowers and fruit produced on the dorsal surface but essentially unknown in Illinois specimens.

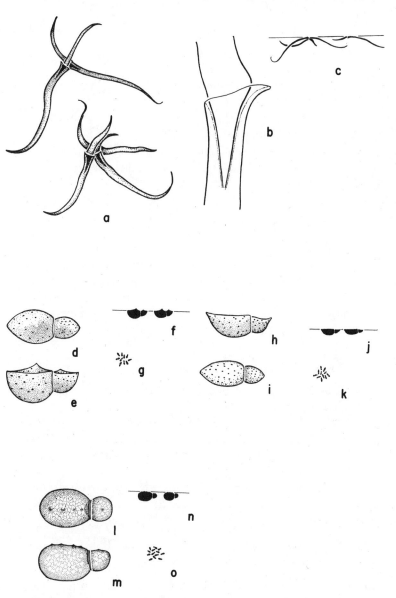

62. *Wolffiella floridana* (Duckweed). *a.* Habit, dorsal aspect, X2½. *b.* Basal reproductive pouch showing vegetative origin of new plant, X10. *c.* Floating habit, lateral aspect, X1¾. *Wolffia papulifera* (Water Meal). *d.* Lateral aspect, habit, X10. *e.* Habit, dorsal aspect, X10. *f.* Floating habit, lateral aspect, X2½. *g.* Floating habit, dorsal aspect, X½. *Wolffia punctata* (Water Meal). *h.* Habit, lateral aspect, X10. *i.* Habit, dorsal aspect, X10. *j.* Floating habit, lateral aspect, X2½. *k.* Floating habit, dorsal aspect, X½. *Wolffia columbiana* (Water Meal). *l.* Habit, lateral aspect, X10. *m.* Habit, dorsal aspect, X10. *n.* Floating habit, lateral aspect, X2½. *o.* Floating habit, dorsal aspect, X½.

COMMON NAME: Duckweed.

HABITAT: Standing water.

RANGE: Massachusetts to Missouri, south to Texas and Florida.

ILLINOIS DISTRIBUTION: Restricted to a few counties in southern Illinois; also Cook County.

The strap-shaped fronds and frequently formed stellate colonies readily distinguish this member of the Lemnaceae from any others in Illinois.

4. *Wolffia* HORKEL

Plants reduced, globoid or ellipsoid, nerveless; upper surface flattened or convex; lower surface strongly convex; pigment cells present in two species; rootlets absent; vegetative reproduction from a single, basal, reproductive pouch; flowers and fruit rarely observed, produced on the upper surface; ovule one.

KEY TO THE SPECIES OF Wolffia IN ILLINOIS

1. Plants punctate, flattened above, with a single papule, or with papule absent.
    2. Upper surface with a single acuminate papule, ovate-elliptic___ _____1. *W. papulifera*
    2. Upper surface flattened, without a papule, elliptic_____ _____2. *W. punctata*
1. Plants epunctate, convex above, frequently with a median row of small papules_____3. *W. columbiana*

1. Wolffia papulifera Thompson, Rep. Mo. Bot. Gard. 9:40. 1898. *Fig. 62* (d–g).

Plants obovoid to ellipsoid, 1.0–1.5 mm long, 0.6–1.0 mm wide, nerveless, cavernous throughout, with pigment cells present; upper surface somewhat flattened but gradually rising in the center to form a single, distinct, conical papule, the apex broadly rounded to subacute, guard cells present; lower surface strongly convex; rootlets absent; plants solitary but usually remaining attached on a common axis in groups of 2; flowers and fruit produced on the upper surface, but rarely observed in Illinois specimens.

COMMON NAME: Water Meal.

HABITAT: Standing water.

RANGE: Eastern United States; Mexico; Argentina.

ILLINOIS DISTRIBUTION: Scattered throughout the state, but apparently not common.

The single, acuminate papule on the upper surface of this species clearly distinguishes it from the other Wolffias in Illinois. *Wolffia papulifera*, as well as *W. punctata*, are extremely difficult to identify in the dried condition. Careful examination under high magnification, however, usually reveals the characteristic surface papule of *W. papulifera*. Both of these plants can be separated from *W. columbiana* on the basis of their pigment cells which remain visible in dried specimens.

2. **Wolffia punctata** Griseb. Fl. Brit. W. Ind. 512. 1864. *Fig. 62* (h–k).

Plants narrowly ovoid to ellipsoid, 0.6–1.0 mm long, 0.3–0.6 mm wide, nerveless, cavernous throughout with pigment cells abundant; upper surface flattened to slightly convex, the apex subacute to acute, the guard cells present; lower surface strongly convex; rootlets absent; plants solitary but usually remaining attached on a common axis in groups of 2; flowers and fruit produced on the upper surface but rarely observed in Illinois specimens.

COMMON NAME: Water Meal.

HABITAT: Standing water.

RANGE: Canada; eastern United States; Mexico; West Indies.

ILLINOIS DISTRIBUTION: Known from four counties in the northern half of Illinois; also Wabash County.

The species is similar to *W. papulifera*, differing essentially by its lack of the conical, surface papule and by its narrower dimensions.

3. **Wolffia columbiana** Karst. Bot. Unters. 1:103. 1865. *Fig. 62* (l–o).

Plants globoid to ellipsoid, 0.5–1.2 mm long, 0.5–0.8 mm wide, nerveless, conspicuously cavernous throughout, with pigment cells lacking; upper surface somewhat flattened, often with a median row of minute papules, guard cells present; lower surface broadly convex; rootlets absent; plants solitary but usually remaining attached on a common axis in groups of 2; flowers and fruits arising on the upper surface but rarely observed in Illinois specimens.

COMMON NAME: Water Meal.

HABITAT: Standing water.

RANGE: Eastern United States; Mexico; Central America; South America.

ILLINOIS DISTRIBUTION: Throughout the state.

Wolffia columbiana, the most widespread species of *Wolffia* in Illinois, differs by its epunctate fronds. Dried specimens become two dimensional but remain easily separated from the other Wolffias.

Order Typhales

Flowers unisexual; perianth parts 0, 3, or 6; stamens 1–7; ovary 1, superior, 1- to 2-celled.

This order is composed of families considered to be highly advanced but appearing simple through reduction of parts. Many classification systems in the past have treated the Typhales as extremely primitive among monocots, a view rejected here.

KEY TO THE FAMILIES OF Typhales IN ILLINOIS

1. Inflorescence axillary, of globose heads; perianth parts 3 or 6; stamens 5, with free filaments_____Sparganiaceae, page 155
1. Inflorescence terminal, elongate; perianth none; stamens 1–7, usually 3, with connate filaments_____Typhaceae, page 163

SPARGANIACEÆ – BUR-REED FAMILY

Only the following genus comprises this family.

1. *Sparganium* L. – Bur-reed

Perennial aquatics from stout rhizomes; leaves alternate, elongate, sheathing at the base; inflorescence globose, axillary; flowers unisexual; perianth parts 3 or 6; stamens 5, with slender, free filaments; ovary 1, superior, 1- to 2-celled, each cell with 1 ovule; fruit a beaked achene subtended by the persistent perianth.

If flowering material only is gathered, care should be taken to observe the number of stigmas in each pistillate flower.

KEY TO THE SPECIES OF Sparganium IN ILLINOIS

1. Plants floating or suberect, to 30 cm long; staminate head 1; pistillate heads about 1 cm in diameter, the lowest short-pedunculate; beak of achene 0.5–1.5 mm long_____1. *S. minimum*
1. Plants erect, (5–) 30–120 cm tall; staminate heads 3–25; pistillate heads 1.5–3.5 cm in diameter, all sessile; beak of achene 2–6 mm long.
   2. Central axis bearing 1–4 pistillate heads; achene stipitate, fusiform, tapering to the summit; stigma 1.
      3. At least one of the pistillate heads borne above the subtending bract (supra-axillary); inflorescence simple; achene usually greenish-brown, even at maturity____2. *S. chlorocarpum*

3. All heads axillary in the subtending bract; inflorescence usually with 1 or more branches; achene pale or dark brown.
    4. Leaves stiff, strongly keeled; branches of inflorescence
without any pistillate heads; pistillate heads 2.5–3.5 cm in
diameter; body of achene 5–7 mm long, shining, pale
brown_____3. *S. androcladum*
    4. Leaves soft, usually not strongly keeled; branches of in-
florescence with 1–3 pistillate heads; pistillate heads
1.5–2.5 cm in diameter; body of achene 3–5 mm long,
dull, dark brown_____4. *S. americanum*
2. Central axis bearing only staminate heads; achene not stipitate,
obpyramidal, truncate at the summit; stigmas 2_____
_____5. *S. eurycarpum*

1. **Sparganium minimum** Fries, Summa Veg. Scand. 2:560.
1849. *Fig. 63.*

Stems floating or suberect, to 30 cm long; leaves to 50 cm long,
2–7 mm wide, thin, without a keel; bracts similar to the leaves,
smaller; inflorescence simple, bearing 1–3 pistillate heads and
1 staminate head, the lowest pistillate head short-pedunculate;
pistillate heads about 1 cm in diameter; sepals spatulate, a little more than half as long as the achene; stigma 1; achene ellip-
tic-ovoid or fusiform, abruptly narrowed at the apex, dull,
greenish or brownish, the body 3–4 mm long, the beak 0.5–1.5
mm long, the stipe 1–2 mm long.

COMMON NAME: Least Bur-reed.
HABITAT: Shallow water.
RANGE: Newfoundland to Alaska, south to California,
Illinois, and New Jersey; Europe.
ILLINOIS DISTRIBUTION: Very rare, known only from a
Vasey collection from McHenry County; not collected
in nearly 100 years and probably extinct in the state.
Since the nearest station of this species to McHenry
County is extremely distant, there may be some doubt
as to the validity of the locality for the Vasey collec-
tion.

2. **Sparganium chlorocarpum** Rydb. N. Am. Fl. 17:8. 1909.
Fig. 64.

Sparganium simplex var. *acaule* Beeby ex Macoun, Cat.
Canad. Pl. 5:367. 1890.

63. *Sparganium minimum* (Least Bur-reed). *a.* Habit, X¼. *b.* Achene, X3¾.

Sparganium acaule (Beeby) Rydb. N. Am. Fl. 17:8. 1909.

Sparganium chlorocarpum var. *acaule* (Beeby) Fern. Rhodora 24:29. 1922.

Stems erect, to 60 cm tall; leaves to 80 cm long, 2–10 mm wide, thin, slightly keeled; bracts similar to the leaves, smaller, ascending to erect; inflorescence simple, bearing 1–4 pistillate heads and 2–9 staminate heads; pistillate heads sessile, 1.5–2.5 cm in diameter; sepals spatulate, about two-thirds as long as the achene; stigma 1; achene fusiform, tapering to the apex, shining, greenish to brownish, the body 4–6 mm long, the beak 2.0–4.5 mm long, the stipe 2.5–3.5 mm long.

COMMON NAME: Green-fruited Bur-reed.

HABITAT: Shallow water.

RANGE: Newfoundland to North Dakota south to Iowa and Virginia.

ILLINOIS DISTRIBUTION: Rare; recorded from only four counties (Cook, Lake, Lee, and Union) widely separated in the state.

After examining specimens of *S. chlorocarpum* and *S. acaule* from the entire range of these plants, it has been concluded that *S. acaule* is merely a smaller phase with the pistillate heads more crowded. It is not sharply defined from *S. chlorocarpum.*

Achenes in *S. chlorocarpum* generally remain greenish, even at maturity. In addition, one or more of the pistillate heads is supra-axillary, that is, borne above the subtending bract.

3. **Sparganium androcladum** (Engelm.) Morong, Bull. Torrey Club 15:78. 1888. *Fig. 65.*

Sparganium simplex var. *androcladum* Engelm. in Gray, Man. Bot. 481. 1867.

Sparganium americanum var. *androcladum* (Engelm.) Fern. & Eames, Rhodora 9:87. 1907.

Sparganium lucidum Fern. & Eames, Rhodora 9:87. 1907.

Stems erect, to about 1 m tall; leaves to 75 cm long, 5–15 mm wide, stiff, keeled; bracts similar to the leaves, smaller; inflorescence usually branched, occasionally simple, the central axis bearing 1–4 pistillate heads and 5–10 staminate heads, the branches without pistillate heads and with 3–6 staminate heads; pistillate heads 2.5–3.5 cm in diameter, sessile; sepals spatulate,

64. *Sparganium chlorocarpum* (Green-fruited Bur-reed). *a.* Habit, X¼.
b. Achene, X2½.

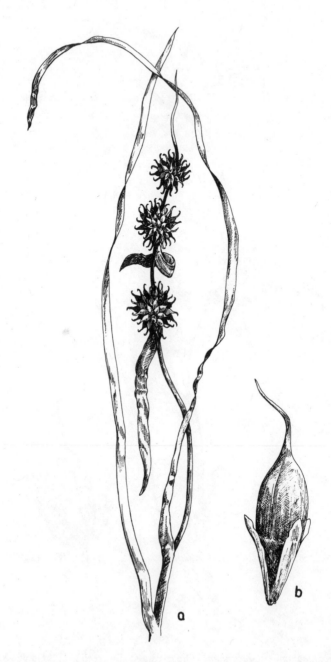

65. *Sparganium androcladum* (Bur-reed). *a.* Habit, X¼.   *b.* Achene, X3.

about two-thirds as long as the achene; stigma 1; achene ellips-oid-fusiform, tapering to the apex, shining, pale brown, the body 5–7 mm long, the beak 4–6 mm long, the stipe 2–4 mm long.

COMMON NAME: Bur-reed.

HABITAT: Shallow water.

RANGE: Quebec to Minnesota, south to Oklahoma and Virginia.

ILLINOIS DISTRIBUTION: Not common; scattered throughout the state.

The flowering heads appear in June and July. The achene which tapers to the summit, and the presence of but a single stigma separate this species from the similar S. eurycarpum. From S. americanum this species differs by its larger, shining, pale brown achenes and its stiffer, keeled leaves.

4. **Sparganium americanum** Nutt. Gen. Pl. 2:203. 1818.
   Fig. 66.

Stems erect, to nearly 1 m tall; leaves to 80 cm long, 4–15 mm wide, thin, usually scarcely keeled; bracts similar to the leaves, smaller; inflorescence usually branched, occasionally simple, the central axis bearing 1–4 pistillate heads and 3–10 staminate heads, the branches bearing 1–3 pistillate heads and 1–6 staminate heads; pistillate heads 1.5–2.5 cm in diameter, sessile; sepals spatulate, about two-thirds as long as the achene; stigma 1; achene fusiform, tapering to the apex, dull, dark brown, the body 3–5 mm long, the beak 2–5 mm long, the stipe 2–3 mm long.

COMMON NAME: Bur-reed.

HABITAT: Shallow water.

RANGE: Newfoundland to Ontario, south to North Dakota, Alabama, and Florida.

ILLINOIS DISTRIBUTION: Not common; restricted to the northern one-fourth of the state.

This species has very obviously thin leaves when compared with the leaves of the similar S. androcladum and S. eurycarpum. It differs further from S. eurycarpum by its single stigma and by the achene which tapers to the summit.

66. *Sparganium americanum* (Bur-reed). *a.* Habit, X¼. *b.* Achene, X2½.

5. **Sparganium eurycarpum** Engelm. in Gray, Man. Bot. 430. 1856. *Fig. 67.*

Stems erect, usually at least 1 m tall; leaves to 75 cm long, 6–12 mm wide, stiff, keeled; bracts similar to the leaves, shorter; inflorescence branched, the central axis bearing only staminate heads, the branches bearing 1–3 pistillate heads and 0–20 staminate heads; pistillate heads 2.0–3.5 cm in diameter, sessile; sepals spatulate, frequently falling early; stigmas 2; achene obpyramidal, truncate at the summit, rather dull, brown, the body 6–9 mm long, the cleft beak 2.5–3.5 mm long, without a stipe.

COMMON NAME: Bur-reed.
HABITAT: Shallow water.
RANGE: Newfoundland to British Columbia, south to California, Oklahoma, and New Jersey.
ILLINOIS DISTRIBUTION: Occasional throughout the state, except for the southeastern counties where it is apparently absent.
This is the largest and most abundant bur-reed in the state. It is readily distinguished by its 2 stigmas and its achene which is abruptly truncate at the summit. The stems attain a greater length and diameter than any other species of *Sparganium* in Illinois.

TYPHACEÆ – CAT-TAIL FAMILY
Only the following genus comprises this family.

1. *Typha* L. – Cat-tail
Perennials from stout rhizomes; leaves cauline, elongated, sheathing at the base; inflorescence terminal, elongated; flowers unisexual; perianth none; stamens 1–7, usually 3, the filaments connate; ovary 1, superior, 1-celled, stipitate, with 1 ovule; achene stipitate, the style persistent.

KEY TO THE SPECIES OF Typha IN ILLINOIS
1. Pollen grains borne in groups of 4; stigma spatulate; leaves flat, usually 8 or more per plant; staminate and pistillate portions of the inflorescence usually contiguous, but not always_____1. *T. latifolia*
1. Pollen grains borne singly; stigma linear; leaves somewhat convex, usually less than 8 per plant; staminate and pistillate portions of the inflorescence separated_____2. *T. angustifolia*

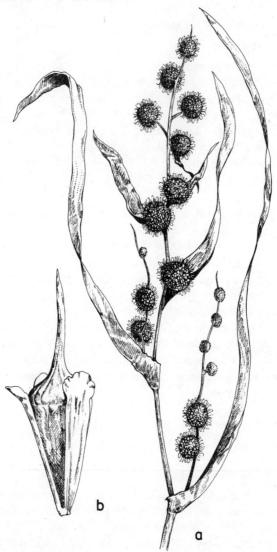

67. *Sparganium eurycarpum* (Bur-reed). *a.* Habit, X³⁄₁₆. *b.* Achene, X2½.

1. **Typha latifolia** L. Sp. Pl. 971. 1753. *Fig. 68.*

Coarse perennial from stout, creeping rhizomes; stems to 4 m tall, with numerous cauline leaves; leaves elongated, flat, 8–22 mm broad, usually 8 or more per plant; staminate and pistillate parts of the inflorescence usually but not always contiguous;

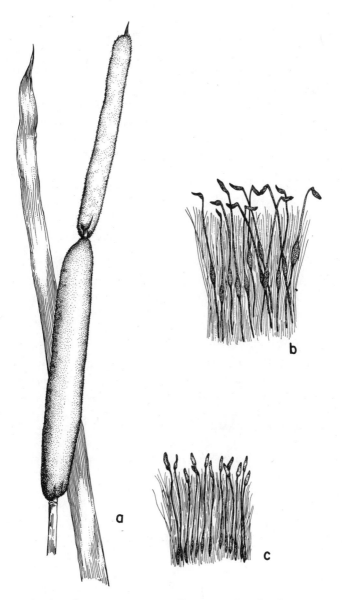

68. *Typha latifolia* (Common Cat-tail). *a.* Leaf and inflorescence, X¼.
b. Pistillate flowers, X3. *c.* Staminate flowers, X3.

staminate inflorescence 5–12 cm long; pollen grains borne in groups of 4; pistillate inflorescence 5–18 cm long, dark brown; stigma spatulate; achene 1 mm long, surrounded by numerous white hairs up to 1 cm long.

COMMON NAME: Common Cat-tail.

HABITAT: Low situations.

RANGE: Throughout most of North America; Europe; Asia.

ILLINOIS DISTRIBUTION: Common; in every county.

Several Illinois specimens of *T. latifolia* have the staminate spikes separated from the pistillate spikes so that this character alone should not be used to distinguish this species from *T. angustifolia*. Neither is leaf width an infallible character, since there is an overlapping between the two species. The number of leaves per plant seems to be as effective as any way in distinguishing the species.

2. **Typha angustifolia** L. Sp. Pl. 971. 1753. *Fig. 69.*

Coarse perennial from stout, creeping rhizomes; stems to 1.5 m tall, with several cauline leaves; leaves elongated, somewhat convex, 4–8 mm wide, less than 8 per plant; staminate and pistillate parts of the inflorescence separated; staminate inflorescence 7–17 cm long; pollen grains borne singly; pistillate inflorescence 4–18 cm long, reddish-brown; stigma linear; achene about 1 mm long, surrounded by numerous white hairs to 8 mm long; 2n = 30 (Gadella & Kliphuis, 1963).

COMMON NAME: Narrow-leaved Cat-tail.

HABITAT: Wet situations.

RANGE: Nova Scotia to Ontario, south to Nebraska, Kentucky, and South Carolina; California; Africa; Europe; Asia.

ILLINOIS DISTRIBUTION: Occasional; scattered throughout the state.

Although once considered rare in Illinois, this species has been found with greater regularity since about 1960.

Hybrids between this and the preceding species have been suspected in Illinois.

69. *Typha angustifolia* (Narrow-leaved Cat-tail). X⅜.

Order Commelinales

Flowers perfect or unisexual; perianth parts 3 or 6 or absent, when present, often distinguishable into 2 series; stamens mostly 1–6; ovary 1, superior; fruit various.

The families comprising this order have been variously treated in the past by many authors; their inclusion into an order here follows the basic principles set forth by Thorne (1963) and others.

KEY TO THE FAMILIES OF Commelinales IN ILLINOIS

1. Perianth present, composed of either calyx or corolla or both.
    2. Calyx and corolla differentiated (in color and texture).
        3. Flowers crowded together in a dense head; leaves basal____
        _____**Xyridaceae, p.** 168
        3. Flowers borne in cymes or umbels; leaves cauline_____
        _____**Commelinaceae, p.** 172
    2. Calyx and corolla undifferentiated (*i.e.*, similar in color and texture).
        4. Perianth petaloid_____**Pontederiaceae, p.** 187
        4. Perianth scarious_____**Juncaceae, p.** 194
1. Perianth absent, or reduced to very minute scales or bristles.
    5. Leaves 2-ranked; sheaths usually open; stems usually hollow with solid nodes, often terete; anthers attached above the base
    _____ **Poaceae***
    5. Leaves 3-ranked; sheaths closed; stems solid, with soft nodes often 3-angled; anthers attached at the base_____**Cyperaceae***
* This family will appear in subsequent volumes in this series.

XYRIDACEÆ – YELLOW-EYED GRASS FAMILY

Perennial herbs with narrow, erect, basal leaves; flowers perfect, crowded in a dense head upon an elongated scape; calyx bilaterally symmetrical, distinguishable in color from the radially symmetrical corolla; fertile stamens 3, attached to the base of the corolla; staminodia present; ovary 1-celled, superior.

Only the following genus occurs in Illinois.

1. *Xyris* L. – Yellow-eyed Grass

Lateral sepals keeled, firm, the anterior sepal not keeled, membranous; corolla yellow; staminodia cleft at apex; capsule ellipsoid, many-seeded.

The family is chiefly one of the Coastal Plain in the United States. Kral (1960) has recently studied this genus.

KEY TO THE SPECIES OF Xyris IN ILLINOIS

1. Plants twisted, from a bulbous base; lateral sepals 3–4 mm long, the keel ciliate_____1. *X. torta*
1. Plants not twisted, from a non-bulbous base; lateral sepals 4–6 mm long, the keel toothed but not ciliate_____2. *X. jupicai*

1. Xyris torta Sm. in Rees, Cyclop. 39. 1818. *Fig. 70.*

Plants bulbous, the bulb up to 1 cm long and thick; leaves 10–30 cm long, 1–2 (–3) mm broad, gray or blue-green, stiff, twisted; scape twisted, bicostate, to 80 cm tall; head ovoid to subgloboid, 6–12 mm long, obtuse at the apex; bracts pale brown, obovate to orbicular, with a gray-green central area; lateral sepals 3–4 mm long, the keel ciliate; seeds 0.5 mm long.

COMMON NAME: Twisted Yellow-eyed Grass.
HABITAT: Low, sandy areas.
RANGE: Maine to Minnesota, south to Texas and Georgia.
ILLINOIS DISTRIBUTION: Not common; restricted to seven counties in the northern half of the state.
The twisted leaves and scapes and the presence of a bulb distinguish this species from the next. The flowers appear in July and August.

This species is primarily one of the Coastal Plain, but also occurs in low, sandy situations from Ohio to Iowa.

2. Xyris jupicai L. Rich. Act. Soc. Hist. Nat. Paris. 1:106. 1792. *Fig. 71.*

Xyris caroliniana Walt. Fl. Carol. 69.1788, nomen confusum.
Plants not bulbous; leaves 10–60 cm long, 1–7 mm broad, greenish, soft, not twisted; scape not twisted, bicostate, to 85 cm tall; head ovoid, 8–15 mm long, obtuse to subacute at the apex; bracts pale brown, with a conspicuous green central area; lateral sepals 4–6 mm long, the keel toothed, not ciliate; seeds 0.5 mm long.

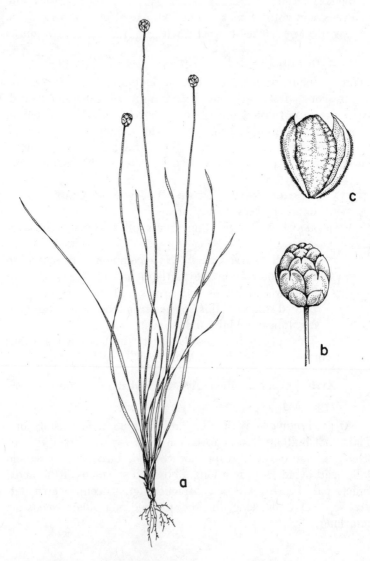

70. *Xyris torta* (Twisted Yellow-eyed Grass).   *a.* Habit, X¼.   *b.* Flowering head, X1½.   *c.* Fruit with bracts, X3.

71. *Xyris jupicai* (Yellow-eyed Grass). *a.* Habit, X½. *b.* Flowering head, X⅜. *c.* Bract, X3¾.

COMMON NAME: Yellow-eyed Grass.

HABITAT: Low, sandy areas.

RANGE: Nova Scotia to Wisconsin, south to Louisiana and Florida.

ILLINOIS DISTRIBUTION: Very rare; exact locality unknown; only collection was made by Gandoger, without any further data other than Illinois. The specimen is in the herbarium of the Missouri Botanical Garden. Since this Coastal Plain species has been found in Porter, Jasper, and LaPorte counties, Indiana, near Lake Michigan, there is reason to believe that Gandoger's collection did come from Illinois.

Kral (1960) has discussed the reasons for rejecting the better known binomial *X. caroliniana* for *X. jupicai.*

COMMELINACEÆ – SPIDERWORT FAMILY

Annuals or perennials from fibrous roots; leaves cauline, sometimes more or less fleshy, parallel-veined, sheathing at the base, the upper frequently modified into a spathe; flowers perfect, cymose or somewhat umbellate; calyx and corolla differentiated, radially or bilaterally symmetrical, ephemeral; fertile stamens 6 (rarely 5), or 3 and with 3 staminodia; ovary 2- to 3-celled, superior; fruit a capsule.

KEY TO THE GENERA OF Commelinaceae IN ILLINOIS

1. Calyx radially symmetrical; corolla radially symmetrical; fertile stamens 6, the filaments villous; ovary 3-celled, with all cells fertile; inflorescence appearing umbellate, from an elongated, leaf-like bract_____1. *Tradescantia*
2. Calyx bilaterally symmetrical, with two of the sepals connate at the base; corolla bilaterally symmetrical; fertile and sterile stamens each 3 (rarely 3 plus 2), the filaments glabrous; ovary 3-celled, with one cell aborted; inflorescence cymose, from a spathe____2. *Commelina*

1. *Tradescantia* L. – Spiderwort

Perennials from fibrous roots; inflorescence cymose but appearing umbellate, borne from the axil of an elongated, leaf-like bract; flowers radially symmetrical; sepals green; corolla colored, ephemeral; fertile stamens 6, the filaments villous; ovary with 3 fertile cells, each with 1–2 ovules; capsule loculicidal, 2- to 6-seeded.

Anderson and Woodson (1935) have made a thorough study of this genus.

The beautiful flowers rapidly disintegrate upon picking. Herbarium specimens fail to show the characters of the corolla properly.

KEY TO THE SPECIES OF Tradescantia IN ILLINOIS

1. Stems flexuous; leaves usually 2–4 cm broad; cymes several, terminal and lateral; sepals 5–10 mm long_____1. *T. subaspera*
1. Stems straight; leaves usually 0.5–1.5 cm broad; cyme usually solitary, terminal; sepals 8–15 mm long.
   2. Stems generally 4–10 dm tall; leaves and stems glabrous and glaucous; sepals glabrous or with tuft of eglandular hairs at the tip_____2. *T. ohiensis*
   2. Stems generally 0.5–4.0 dm tall; leaves and stems glabrous or pubescent, but not glaucous; sepals pubescent throughout with glandular or eglandular hairs.
      3. Pubescence of sepals eglandular_____3. *T. virginiana*
      3. Pubescence of sepals glandular-viscid_____4. *T. bracteata*

1. **Tradescantia subaspera** Ker, Bot. Mag. 39:1597. 1813. *Fig. 72.*

Tradescantia pilosa Lehm. Ind. Sem. Hort. Hamb. 16. 1827.
Tradescantia subaspera var. *typica* Anders. & Woodson, Contr. Arnold Arb. 9:49. 1935.
Stems stout, flexuous, green, glabrous or sparsely pubescent, 4–10 dm tall; leaves dark green, firm, broadly lanceolate, acute to acuminate, tapering to a slender petiole-like base, to 20 cm long, 2–4 cm wide, glabrous or sparsely pilose; cymes several, terminal and lateral; pedicels 10–20 mm long, glabrous or pubescent; sepals 5–10 mm long, pubescent, with glandular and eglandular hairs; petals 10–15 mm long, blue; capsule 4–6 mm long; seeds 2–3 mm long.

COMMON NAME: Spiderwort.
HABITAT: Woodlands.
RANGE: West Virginia to Illinois, south to Missouri and Tennessee.
ILLINOIS DISTRIBUTION: Occasional in the southern two-thirds of Illinois; rare in the northern one-third.
This handsome species flowers from the last of May to early August. It comes into flower later than any other *Tradescantia* in Illinois.
Tradescantia subaspera has the broadest leaves of any

72. *Tradescantia subaspera* (Spiderwort).   *a.* Habit, X¼.   *b.* Capsule, X2.
c. Seed, X4.

Tradescantia in the state. Variation occurs as to the amount of pubescence on all parts of the plant. The hairs may be glandular or eglandular.

Until 1940, this plant in Illinois had been known as *T. pilosa*, but this name is not the earliest applicable to this species.

2. **Tradescantia ohiensis** Raf. Precis Decouv. Somiol. 45. 1814. *Fig. 73.*

Tradescantia canaliculata Raf. Atl. Journ. 1:150. 1832.
Tradescantia reflexa Raf. New Fl. N. Am. 2:87. 1837.

Stems slender, usually 4–10 dm tall, straight, glaucous, glabrous; leaves glaucous, firm, linear-lanceolate, acuminate, to 40 cm long, rarely over 1 cm broad, the blade glabrous, the sheath glabrous or puberulent; cyme usually one, terminal; pedicels 8–22 mm long, glabrous; sepals 8–14 mm long, glabrous or with a tuft of eglandular hairs at the tip; petals 12–20 mm long, blue (rarely rose or white); capsule to 3.5 mm long; seeds light gray, to 2 mm long.

COMMON NAME: Spiderwort.
HABITAT: Edges of woodlands, prairies.
RANGE: Massachusetts to Minnesota, south to Texas and Florida.
ILLINOIS DISTRIBUTION: Rather common; probably in every county but thus far not collected in every county.
This species usually flowers from late April to late August. The glaucous stems and leaves and the rather large stature of the plants make it easily distinguishable. The white-flowered f. *albiflora* has been collected in Bond County (*Slagle 8*).

In addition to the binomial *T. ohiensis*, Rafinesque gave two other names to plants referable to this species, and both of these names were frequently used by Illinois botanists until about 1950.

3. **Tradescantia virginiana** L. Sp. Pl. 288. 1753. *Fig. 74.*

Tradescantia virginica L. Syst. 10:975. 1759.
Tradescantia brevicaulis Raf. Atl. Journ. 1:150. 1832.
Tradescantia virginica var. *villosa* S. Wats. Proc. Am. Acad. n.s. 18:168. 1883.

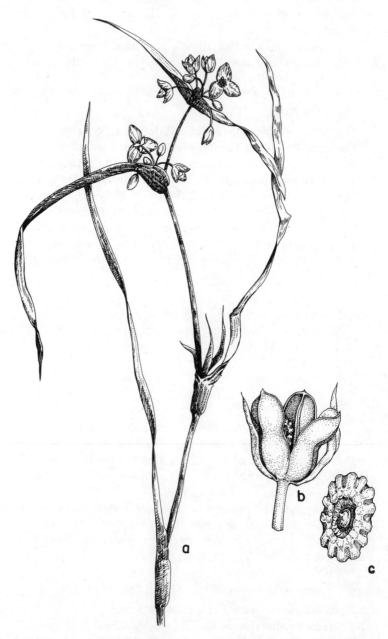

73. *Tradescantia ohiensis* (Spiderwort). *a.* Habit, X¼. *b.* Dehiscing fruit, X2. *c.* Seed, X4.

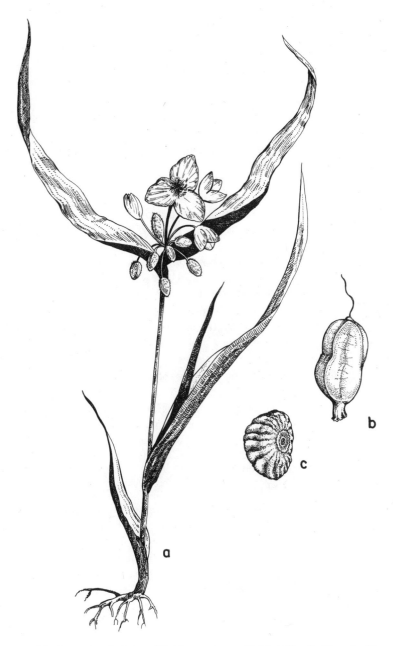

74. Tradescantia virginiana (Spiderwort). *a.* Habit, X¼. *b.* Capsule, X2.
c. Seed, X3¾.

Tradescantia virginica f. *albiflora* Britt. Bull. Torrey Club 17:125. 1890.

Stems rather slender, 0.5–4.0 dm tall, straight, green, glabrous or puberulent; leaves green, thin, linear-lanceolate, long-acuminate, to 30 cm long, 5–15 mm wide, glabrous or nearly so; cyme usually one, terminal; pedicels 15–35 mm long, sparsely pilose with eglandular hairs; sepals 10–15 mm long, puberulent throughout with eglandular hairs; petals 12–20 mm long, basically blue, but varying from rose to purple to white; capsule 4–7 mm long; seeds 2–3 mm long; 2n = 24 (Darlington & Vosa, 1963).

COMMON NAME: Spiderwort.

HABITAT: Woodlands and prairies.

RANGE: Maine to Minnesota, south to Missouri and Georgia.

ILLINOIS DISTRIBUTION: Common in the southern two-thirds of the state; rare in the northern one-third.

Flower color may be virtually any shade of blue, purple, or violet. Beautiful rose-colored flowers are known, as are white flowers. Flowering time is late April to early June.

This and the following species are very similar, differing mainly in the presence or absence of glandular hairs on the pedicels and sepals.

This species was known as *T. virginica* until Anderson and Woodson (1935) pointed out that Linnaeus had called it originally *T. virginiana.*

4. **Tradescantia bracteata** Small in. Britt. & Brown, Ill. Fl. 3:510. 1898. *Fig. 75.*

Stems rather slender, 2–4 dm tall, straight, green, glabrous or puberulent; leaves green, rather firm, linear-lanceolate, long-acuminate, to 35 cm long, 8–15 mm wide, glabrous or pubescent near the base; cyme usually one, terminal; pedicels 10–20 mm long, with glandular-viscid hairs; sepals 10–15 mm long, glandular-viscid; petals 15–20 mm long, rose; capsule 4–7 mm long; seeds to 2.5 mm long, tan to dark gray; 2n = 12 (Darlington & Vosa, 1963).

75. *Tradescantia bracteata* (Prairie Spiderwort).   *a*. Habit, X³⁄₁₆.   *b*. Capsule, X2.   *c*. Seed, X3¾.

COMMON NAME: Prairie Spiderwort.

HABITAT: Prairies.

RANGE: Michigan to Montana, south to Kansas and Indiana.

ILLINOIS DISTRIBUTION: Very rare; known from three counties in the central one-third of the state.

The solitary, terminal cyme and the size of the floral parts between this species and *T. virginiana* are similar.

2. *Commelina* L. – Day Flower

Annuals or perennials from fibrous roots or rhizomes; inflorescence cymose, borne from a spathe; flowers bilaterally symmetrical; sepals green, with two of them basally connate; petals colored or white, unequal in size, ephemeral; fertile stamens 3, with two of them erect and one curved inward; sterile stamens 3 (rarely 2), with cruciform anthers; filaments glabrous; ovary with 2 fertile and 1 sterile cell, the fertile each with 1–2 ovules; capsule loculicidal, 2- to 6-seeded.

KEY TO THE SPECIES OF Commelina IN ILLINOIS

1. Plants rooting at the nodes; margins of spathes free; seeds black, rugose or reticulate.
    2. Plants annual; leaf-sheaths glabrous at summit; anthers 6; seeds 4; petals 10–15 mm long, the lower petal white; seeds 3.5–4.0 mm long, rugose_____1. *C. communis*
    2. Plants perennial; leaf-sheaths ciliate at summit; anthers 5; seeds 5; petals 6–8 mm long, the lower petal blue; seeds 2.0–2.5 mm long, reticulate_____2. *C. diffusa*
1. Plants not rooting at the nodes; margins of spathes united for about one-third the way up from the base; seeds reddish or brown, smooth or puberulent, not rugose or reticulate.
    3. Plants with rhizomes; lowest petal blue, only slightly smaller than the other two; seeds about 6 mm long, reddish_____
    _____3. *C. virginica*
    3. Plants with thick, fleshy roots; lowest petal white, much smaller than the other two; seeds about 3 mm long, brown__4. *C. erecta*

1. Commelina communis L. Sp. Pl. 40. 1753. *Fig. 76.*

Annual with slender, fibrous roots; stems erect or decumbent and rooting at the nodes, to 75 cm long; leaves ovate-lanceolate to lanceolate, to 10 cm long, 1–3 cm broad, the sheaths glabrous

76. *Commelina communis* (Common Day Flower). *a.* Habit, X¼. *b.* Spathe, X¾. *c.* Spathe (diagrammatic, with one-half cut away), X¾. *d.* Seed, X3¾.

at the summit; spathe 15–25 mm long, about half as broad when folded, acute or short-acuminate, glabrous or puberulent, the margins free; upper petals blue, 10–15 mm long, the lower petal white, much smaller; anthers 6; seeds 4, 3.5–4.0 mm long, black, rugose.

COMMON NAME: Common Day Flower.

HABITAT: Moist waste ground.

RANGE: Native to Asia; established throughout the eastern United States.

ILLINOIS DISTRIBUTION: Occasional; throughout the state.

This species, which is the only annual day flower in the state, blooms from late June to mid-October. The reduced, lower white petal is like the condition in *C. erecta*. The rugose seeds, however, easily distinguish *C. communis*.

2. **Commelina diffusa** Burm. f. Fl. Ind. 18, pl. 7, f.2. 1768. *Fig. 77.*

Commelina cayennensis Rich. Act. Soc. Hist. Nat. Paris 1:106. 1792.

Commelina agraria Kunth, Enum. Pl. 4:38. 1843.

Perennial from fibrous roots; stems decumbent, rooting at the nodes, to 1 m long; leaves lanceolate, to 8 cm long, 1–2 cm broad, the sheaths long-ciliate at the summit; spathe 12–25 mm long, about half as broad when folded, acute or short-acuminate, glabrous or ciliolate, the margins free; all petals blue, generally similar in size, 6–8 mm long; anthers 5; seeds 5, 2.0–2.5 mm long, black, reticulate.

COMMON NAME: Day Flower.

HABITAT: Moist woodlands and moist waste ground.

RANGE: Kansas to Virginia, south into the American tropics; Asia.

ILLINOIS DISTRIBUTION: Occasional in the southern one-third of the state; absent in the northern two-thirds, except for Jersey and Mason counties.

This is the only species of day flower in Illinois in which all the petals are less than 1 cm long. The presence of five anthers and the reticulate seeds are

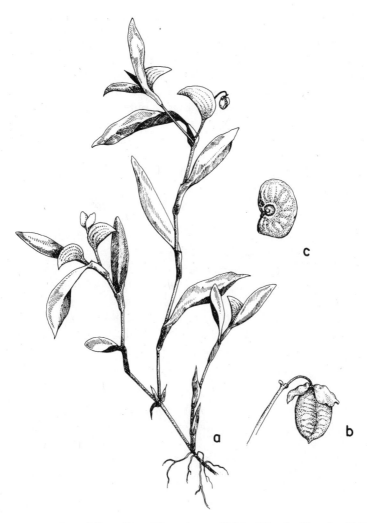

77. *Commelina diffusa* (Day Flower). *a*. Habit, X¼. *b*. Capsule, X1½.
c. Seed, X3¾.

also distinctive. The flowers are produced from July to October.

For many years this species was confused with and often called *C. virginica*.

3. **Commelina virginica** L. Sp. Pl. 61. 1762. *Fig. 78.*

Commelina hirtella Vahl, Enum. Pl. 2:166. 1806.

Perennial from forking rhizomes; stems usually erect, seldom exceeding 1 m tall; leaves lanceolate, to 15 (–20) cm long, 2–5 cm broad, the sheaths reddish-setose at the summit; spathe 15–25 mm long, more than half as wide when folded, short-acuminate, the margins connate for about one-third of the length from the base; all petals blue, similar in size, over 10 mm long; seeds 5, about 6 mm long, reddish, glabrous or puberulent, neither rugose nor reticulate.

COMMON NAME: Day Flower.

HABITAT: Low woodlands.

RANGE: New Jersey to Kansas, south to Texas and Florida.

ILLINOIS DISTRIBUTION: Not common; restricted to the southern one-fourth of the state.

This species flowers from mid-June to early September. The seeds are the largest in this genus for Illinois species. It is the only Illinois species with rhizomes.

Commelina virginica is encountered less frequently than any other species of the genus in Illinois.

4. **Commelina erecta** L. Sp. Pl. 41. 1753. *Fig. 79.*

Commelina angustifolia Michx. Fl. Bir. Am. 1:24. 1803.

Commelina crispa Woot. Bull. Torrey Club 25:451. 1898.

Commelina erecta var. *angustifolia* (Michx.) Fern. Rhodora 42:439. 1940.

Commelina erecta var. *angustifolia* f. *crispa* (Woot.) Fern. Rhodora 42:440. 1940.

Commelina erecta var. *deamiana* Fern. Rhodora 42:449. 1940.

Perennial from fleshy roots; stems arching or erect, rarely exceeding 1 m tall; leaves linear-lanceolate to lanceolate to ovate-lanceolate, to 15 cm long, 0.5–4.0 cm broad, the sheaths glabrous or puberulent and with pale cilia at the summit; spathe 10–35 mm long, short-acuminate, glabrous or pubescent, the margins connate for about one-third of the length from the base; upper petals blue, (10–) 15–25 mm long, the lower petal white, smaller; seeds about 3 mm long, brown, glabrous, neither rugose nor reticulate.

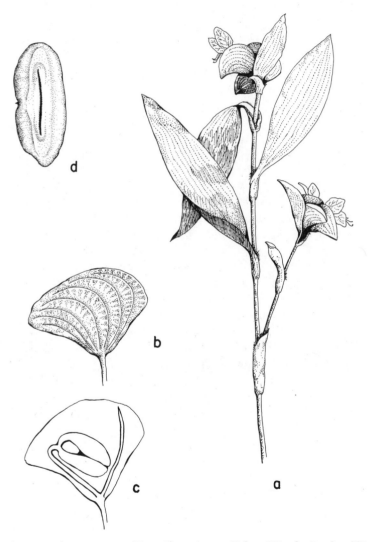

78. *Commelina virginica* (Day Flower). *a*. Habit, X¼. *b*. Spathe, X¾.
c. Spathe (diagrammatic, with one-half cut away), X¾. *d*. Seed, X3¾.

79. *Commelina erecta* (Day Flower). *a*. Habit, X¼. *b*. Flower, X½. *c*. Spathe, X⅝. *d*. Spathe (diagrammatic, with one-half cut away), X⅝. *e*. Seed, X2½.

COMMON NAME: Day Flower.

HABITAT: Moist or dry sandy soil.

RANGE: New York to Wyoming, south to New Mexico and Florida; Mexico; Cuba.

ILLINOIS DISTRIBUTION: Occasional in the western counties; absent from the eastern half of the state, except for Cook and Kankakee counties.

This species is highly variable in stem length, leaf shape and size, and spathe size and pubescence. All varieties recognized by Fernald (1940; 1950), on the basis of Illinois material, overlap in every character. Variety *erecta* generally has taller stems, broader leaves, and longer spathes than varieties *angustifolia* and *deamiana*. Variety *angustifolia* usually has shorter leaves and spathes than var. *deamiana*. No distinction can be made accurately with the Illinois specimens.

PONTEDERIACEÆ–PICKERELWEED FAMILY

Perennial herbs of wet situations from rhizomes; leaves basal and cauline; flowers perfect, borne singly or in spikes or spike-like panicles; perianth parts 6, united below, of uniform color; stamens 3 or 6; ovary 1- or 3-celled, superior; fruit a capsule or utricle.

KEY TO THE GENERA OF Pontederiaceae IN ILLINOIS

1. Perianth funnelform, bilaterally symmetrical; stamens 6; ovary 3-celled, with 2 cells abortive; fruit a utricle; inflorescence a spike-like panicle with more than 8 flowers; perianth segments blue, with the uppermost marked with yellow_____1. *Pontederia*
1. Perianth salverform, radially symmetrical; stamens 3; ovary 1-celled or partially 3-celled; fruit a capsule; inflorescence a spike of 2–8 flowers, or flower solitary; perianth segments blue, white, or pale yellow, but not blue marked with yellow.
   2. Flower pale yellow; stamens uniform; ovary 1-celled; capsule indehiscent, few-seeded; leaves linear, 2–6 mm broad_____ _____2. *Zosterella*
   2. Flowers blue or white; two stamens shorter than the third; ovary partially 3-celled; capsule dehiscent, many-seeded; leaves broader, at least 1 cm broad_____3. *Heteranthera*

1. *Pontederia* L. – Pickerelweed

Perennial from creeping rhizome; leaves basal and cauline, petiolate; inflorescence a spike-like panicle with more than 8 flowers; flowers perfect; perianth funnelform, 2-lipped, with 6 segments free above, blue, with the uppermost segment marked with yellow; stamens 6, 3 of them exserted and fertile, 3 of them included and sometimes sterile; ovary 3-celled; fruit a utricle.

Only the following handsome species occurs in Illinois.

1. Pontederia cordata L. Sp. Pl. 288. 1753. *Fig. 80.*

Pontederia angustifolia Pursh, Fl. Am. Sept. 1:224. 1814.
Pontederia cordata var. *angustifolia* (Pursh) Torr. Fl. U. S. 1:343. 1824.
Pontederia cordata f. *angustifolia* (Pursh) Solms-Laub. in DC. Monog. Phan. 4:532. 1883.

Rhizome thick, widely creeping; lowest stem-leaf and basal leaves similar, ovate to narrowly lanceolate, obtuse to acute at the apex, cordate or tapering to the base, to 20 cm long; petiole rather stout, to 7 cm long; inflorescence subtended by a bladeless sheath, with over 8 flowers in a spike-like panicle; flowers more or less villous; perianth purple, the tube 5–7 mm long, the lobes 6–10 mm long; fruit toothed, beaked, enclosed by the persistent perianth tube, to 1 cm long; seeds 3.5–4.5 mm long; 2n = 16 (Bowden, 1945).

COMMON NAME: Pickerelweed.

HABITAT: Wet situations, sometimes in standing water.

RANGE: Nova Scotia to Ontario, south to Texas and Florida.

ILLINOIS DISTRIBUTION: Occasional; absent from the east-central counties.

The beautiful inflorescence of purple flowers is produced from the last of May until early September. Considerable variation exists in leaf shape, with the range extending from broadly ovate to narrowly lanceolate. There is no apparent line of demarcation to separate this species into forms.

We have observed this species in standing water as well as in shoreline mud. It is most frequent around ponds and lakes, less frequent along rivers and streams.

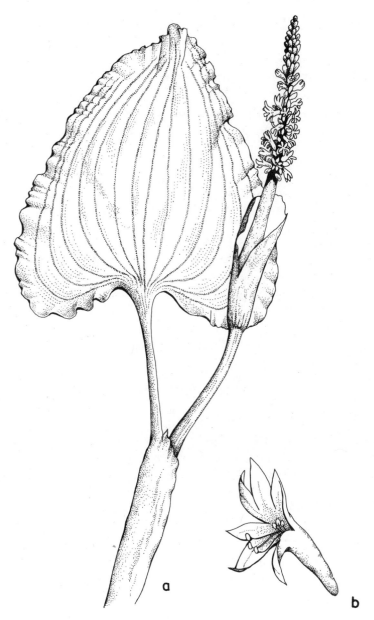

80. *Pontederia cordata* (Pickerelweed). *a.* Inflorescence and leaf, X¼. *b.* Flower, X2.

2. Zosterella SMALL – Water Star Grass

Perennial from rhizomes; leaves cauline, linear; flower solitary, borne from a spathe; perianth salverform, regular; stamens 3, uniform; ovary 1-celled; fruit a few-seeded, indehiscent capsule. Only the following species occurs in Illinois.

1. **Zosterella dubia** (Jacq.) Small in Small & Carter, Fl. Lanc. Co. 68. 1913. *Fig. 81.*

Commelina dubia Jacq. Obs. Bot. 3:9. 1768.
Schollera graminifolia Willd. Neue Schr. Ges. Naturf. Fr. Berlin 3:438. 1801.
Leptanthus gramineus Michx. Fl. Bor. Am. 1:25. 1803.
Heteranthera graminea (Michx.) Vahl, Enum. 2:45. 1805.
Schollera graminea (Michx.) Raf. Am. Mo. Mag. 2:175. 1818.
Heteranthera dubia (Jacq.) MacM. Metasp. Minn. Valley 138. 1892.

Usually aquatic perennial from rather slender rhizomes; leaves linear, obtuse to subacute, to 15 cm long, to 6 mm broad, translucent; flower pale yellow, solitary, borne from a spathe 2–5 cm long; perianth tube to 6 cm long, very slender; capsule ovoid, 8–10 mm long; 2n = 30 (Bowden, 1945).

COMMON NAME: Water Star Grass.
HABITAT: Shallow water; muddy shores.
RANGE: Quebec to Washington, south to New Mexico, Texas, and Florida; California; Mexico.
ILLINOIS DISTRIBUTION: Not common; confined to the northern one-half of the state.

This species usually is placed in the genus *Heteranthera*, but there seems to be real justification in following Small by segregating it into a separate genus. The major differences between *Zosterella* and *Heteranthera* are summarized below.

Zosterella	Heteranthera
Stamens uniform	Stamens of two kinds
Ovary 1-celled	Ovary partially 3-celled
Fruit indehiscent	Fruit dehiscent
Fruit few-seeded	Fruit many-seeded
Leaves linear	Leaves lanceolate
	to orbicular

81. Zosterella dubia (Water Star Grass). *a.* Habit (left), X⅙. *b.* Habit, with flower (right), X¾.

In addition to the generic differences listed above, *Zosterella dubia* differs from the Illinois species of *Heteranthera* by its yellow flowers.

The floral spathe in *Zosterella dubia* caused early workers to place this species in the genus *Commelina*.

This species bears its flowers during July and August.

3. *Heteranthera* R. & P. – Mud Plantain

Perennial from rhizomes; leaves cauline, lanceolate to orbicular; flower solitary, or 2–8 in a spike, borne from a bladeless spathe; perianth salverform, regular; stamens 3, two of them short and with ovate anthers, one of them longer and with a sagittate anther; ovary partially 3-celled; fruit a many-seeded, dehiscent capsule.

KEY TO THE SPECIES OF Heteranthera IN ILLINOIS

1. Flower solitary; perianth tube 2–4 cm long; filaments glabrous; stigma 3-lobed; leaves lanceolate to ovate, tapering, rounded, or subcordate at the base_____1. *H. limosa*
1. Flowers 2–8 in a spike; perianth tube 5–10 cm long; filaments pilose; stigma capitate; leaves orbicular, cordate__2. *H. reniformis*

1. Heteranthera limosa (Sw.) Willd. Neue Schr. Ges. Naturf. Fr. Berlin 3:439. 1801. *Fig. 82.*

Pontederia limosa Sw. Prodr. 57. 1788.

Leptanthus ovalis Michx. Fl. Bor. Am. 1:25. 1803.

Perennial from rhizomes; leaves lanceolate to ovate, obtuse or subacute at the apex, tapering, rounded, or subcordate at the base, to 5 cm long, to 4 cm broad, usually much smaller, with a petiole to 17 cm long; flower solitary, blue marked with one or more white spots, borne from a spathe; spathe long-acuminate, 2–4 cm long; perianth tube 2–4 cm long, the segments linear-lanceolate, all the same width; filaments glabrous; stigma 3-lobed; capsule elongated, 12–15 mm long.

82. *Heteranthera limosa* (Mud Plantain). *a.* Habit, X⅛. *b.* Capsule with spathe, X1¼.

COMMON NAME: Mud Plantain.

HABITAT: Muddy shores; shallow water.

RANGE: Minnesota to Colorado, south to New Mexico and Florida; Mexico; West Indies; Central America; South America.

ILLINOIS DISTRIBUTION: Rare; known only from Alexander, Hardin and St. Clair counties.

The solitary, axillary flower relates this species to *Zosterella dubia,* but the generic differences outlined under that species easily distinguish the two.

Several collections have been made of *Heteranthera limosa* around East St. Louis in St. Clair County since the original Illinois collection in 1838 by George Engelmann. In 1964, Fore and Stookey collected this species from the edge of a farm pond near Elizabethtown in Hardin County. It was discovered along the edge of a fallow field at Horseshoe Lake, Alexander County, in 1968.

The flowers appear in June and July.

2. Heteranthera reniformis R. & P. Fl. Per. 1:43. 1798. *Fig. 83.*

Perennial from rhizomes; leaves orbicular, subacute at the apex, cordate at the base, to 5 cm long, nearly as broad, with a petiole to 15 cm long; inflorescence spicate, 2- to 8-flowered, the flowers pale blue or whitish; spathe short-acuminate, 1–3 cm long; perianth tube 5–10 cm long, the inner segments linear-lanceolate, the outer segments narrower; filaments pilose; stigma capitate; capsule oblongoid, 5–9 mm long.

COMMON NAME: Mud Plantain.

HABITAT: Muddy shores; shallow water.

RANGE: Connecticut to Nebraska, south to Texas and Florida; West Indies; Mexico; Central America; South America.

ILLINOIS DISTRIBUTION: Rare; known only from five counties in the southern one-fourth of the state.

Heteranthera reniformis is distinguished from *H. limosa* by its spicate inflorescence, its elongate perianth tube, its hairy filaments, and its shorter capsules.

This species is restricted to the southern one-fourth of Illinois where it occurs in very shallow water or in shoreline mud.

The flowers are produced in July and August.

JUNCACEÆ–RUSH FAMILY

Annuals or usually perennials, with or without stolons; inflorescence paniculate, cymose, or umbellate; perianth 6-parted, green or brown; stamens 3 or 6; ovary superior, 1- to 3-locular; fruit a capsule; seeds often carunculate.

This family is distinguished from the often somewhat similar appearing Cyperaceae (sedges) and Poaceae (grasses) by the presence of a perianth.

83. Heteranthera reniformis (Mud Plantain). *a.* Habit, X⅙. *b.* Capsule, X1¼.

KEY TO THE GENERA OF Juncaceae IN ILLINOIS

1. Plants hairy; capsule 3-seeded_____1. *Luzula*
1. Plants glabrous; capsule several-seeded_____2. *Juncus*

1. *Luzula* DC. – Wood Rush

Cespitose perennials with or without stolons; leaves pilose; inflorescence an umbel; perianth undifferentiated, 6-parted; stamens 6; ovary superior, 1-locular; capsule 3-seeded, the seeds bearing a caruncle.

KEY TO THE SPECIES OF Luzula IN ILLINOIS

1. Flowers solitary (rarely paired) at the tips of the inflorescence rays; seeds over 2 mm long, including the strongly curved caruncle_____1. *L. acuminata*
1. Flowers crowded in glomerulate spikes; seeds 1.2–2.0 mm long, including the conical caruncle_____2. *L. multiflora*

1. Luzula acuminata Raf. Aut. Bot. 193. 1840. *Fig. 84.*

Luzula saltuensis Fern. Rhodora 5:195. 1903.

Loosely cespitose perennial with occasional stolons; stems to 35 cm tall; leaves broadly linear, cuspidate at the apex, to 3 cm long, to 1.2 cm broad, pilose; inflorescence umbellate, simple, the rays ascending to reflexed, 1- (2-) flowered; perianth segments narrowly ovate, 2.5–4.5 mm long, rich brown with hyaline margins; capsule ovoid, 3.0–4.5 mm long; seeds 1.0–1.5 mm long, with a strongly curved caruncle 1.0–1.5 mm long, purple-brown.

COMMON NAME: Wood Rush.

HABITAT: Open woods.

RANGE: Newfoundland to Saskatchewan, south to South Dakota, Illinois, and Georgia.

ILLINOIS DISTRIBUTION: Known only from Ogle and LaSalle counties.

The single-flowered rays of the simple inflorescence readily distinguish this *Luzula* from any others in Illinois. These rays may be ascending, spreading, or reflexed. This species flowers in early May in Illinois.

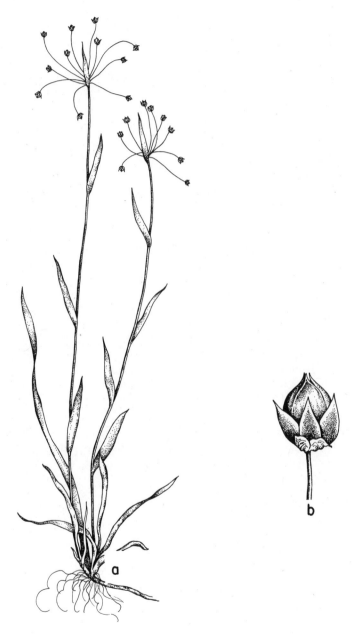

84. Luzula acuminata (Wood Rush). *a.* Habit, X⅛. *b.* Capsule, X3.

2. **Luzula multiflora** (Retz.) Lejeune, Fl. Envir. Spa. 1:169. 1811.

Densely cespitose perennial with occasional white bulb-like tubers borne at the base; stems to 60 cm tall; leaves lance-linear, to 7 mm broad, ciliate and pilose; inflorescence umbellate, the central spike often sessile, the 1–20 other spikes on erect, ascending, spreading, or reflexed rays; spikes 4–12 mm long, 5–9 mm broad; perianth segments lanceolate, acute to acuminate, 2–4 mm long, pale to dark brown with hyaline margins; capsule obovoid, 2.5–4.0 mm long; seeds 1.2–2.0 mm long, with a conical caruncle to 0.7 mm long.

Highly variable species with two recognizable varieties occurring in Illinois.

1. Rays of umbel erect to ascending___2a. *L. multiflora* var. *multiflora*
1. Rays of umbel horizontally spreading to reflexed_____
_____2b. *L. multiflora* var. *echinata*

2a. **Luzula multiflora** (Retz.) Lejeune var. **multiflora** *Fig. 85.*

Juncus multiflorus Retz. Fl. Scand. Prodr. 82. 1795.
Luzula campestris var. *bulbosa* Wood, Class-Book 723. 1861.
Luzula campestris var. *multiflora* (Retz.) Celak, Prodr. Fl. Boehem. 85. 1869.
Luzula bulbosa (Wood) Rydb. Brittonia 1:85. 1931.

White bulb-like tubers occasionally present at base of plant; spikes of umbel 1–20, the central one usually sessile, the others on erect or ascending rays; spikes 4–12 mm long, 5–9 mm broad; perianth segments 2.5–3.5 mm long; seeds 1.2–2.0 mm long.

COMMON NAME: Wood Rush.
HABITAT: Woods and shaded cliffs.
RANGE: Greenland to Alaska, south to California, Texas, and Georgia.
ILLINOIS DISTRIBUTION: Rather common in the southern counties, less common in the northern counties, nearly absent in the central counties.

Some specimens produce white bulb-like tubers at the base. These have been called *L. bulbosa*. Gleason and Cronquist (1963) attribute bulb-bearing forms to both *L. multiflora* and *L. echinata*, but I have never seen bulbs on the latter.

85. *Luzula multiflora* var. *multiflora* (Wood Rush). *a.* Habit, X¼. *b.* Spike, X1⅞. *c.* Capsule, X3. *d.* Seed, X6⅛.

Gleason and Cronquist (1963) call this variety and the following varieties of *L. campestris*. This latter European species
is stolon-bearing with subglobose spikes.

2b. **Luzula multiflora** (Retz.) Lejeune var. **echinata** (Small)
Mohl., stat. nov. *Fig. 86.*

Juncoides echinatum Small, Torreya 1:74. 1901.
Luzula campestris var. *echinata* (Small) Fern. & Wieg. Rho-
dora 15:42. 1913.
Luzula echinata (Small) Hermann, Rhodora 40:84. 1938.
Luzula echinata var. *mesochorea* Hermann, Rhodora 40:84.
1938.

Bulblets absent; spikes of umbel 2–16, the central one usually
sessile, the others on horizontally spreading or reflexed rays;
spikes 6–10 mm long, about as broad; perianth segments 2.5–4.0
mm long; seeds 1.2–1.6 mm long.

HABITAT: Woods and cliffs.
RANGE: Massachusetts to Illinois, south to Texas and
Georgia.
ILLINOIS DISTRIBUTION: Generally confined to the southern half of Illinois; found regularly along the Shawnee-
town Ridge.
Some workers treat *L. multiflora*, *L. bulbosa*, and *L.
echinata* as distinct species. The only reliable difference is in the arrangement of the rays of the umbel,
a character scarcely justifying specific differentiation.
Variety *mesochorea*, considered synonymous here with var.
echinata, seemingly is only a smaller phase of var. *echinata*.

2. *Juncus* L. – Rush

Perennials, rarely annuals; sheath open, often auriculate; blades
flat, involute, terete, or absent; inflorescence paniculate or cymose, bracteate; perianth 6-parted, scarious-margined, green or
brown; stamens 3 or 6; ovary 3-celled or incompletely 3-celled;
capsule with numerous seeds, the seeds apiculate or caudate.

Leafy proliferations in the inflorescences are not unusual in
the septate-leaved species. They are formed when the heads
are pendent in water (Fernald, 1950, with reference to *J. acuminatus*) or through the activity of the homopteran insect *Livia
juncorum* Lat. (Curtis, 1862).

86. *Luzula multiflora* var. *echinata* (Wood Rush). *a.* Habit, X½. *b.* Capsule, X6¼.

Engelmann (1866, 1868) revised the North American species of *Juncus* and later Buchenaú (1906) monographed the genus. The Illinois species have been studied by DeFilipps (1964, 1966).

KEY TO THE TAXA OF Juncus IN ILLINOIS

1. Leaf sheaths without blades, apiculate or mucronate; inflorescences appearing lateral on the stems.
    2. Stems densely cespitose; stamens 3, anthers 0.5–0.8 mm long; seeds 0.5 mm long; capsules beakless___1. *J. effusus* var. *solutus*
    2. Stems single at intervals from elongate rhizomes; stamens 6, anthers 1.5 to nearly 2.0 mm long; seeds 1 mm long; capsules with beaks 0.5–1.0 mm long_____2. *J. balticus* var. *littoralis*
1. Leaf sheaths with definite blades; inflorescences terminal.
    3. Flowers in heads, not prophyllate (*i.e.*, not with bractlets).
        4. Leaves flat, not terete and cross-septate; anthers purplish-brown.
            5. Stems solitary, approximately 3.5 cm apart on conspicuous, scaly rhizomes, 4.9–8.8 (–10.2) dm tall; leaves 2.0–6.5 mm wide; heads (13–) 20–135._____3. *J. biflorus*
            5. Stems cespitose, 0.45–5.90 dm tall; leaves 1.0–2.5 (–2.9) mm wide; heads 3–28 (–32)_____4. *J. marginatus*
        4. Leaves terete, cross-septate; anthers yellow.
            6. Seeds fusiform, 0.8–1.9 mm long, caudate.
                7. Seeds 1.2–1.9 mm long, the tails comprising (one-third) one-half to five-eighths the total length of the seeds; heads 5- to 50-flowered; stamens 3_____ _____5. *J. canadensis*
                7. Seeds 0.8–1.2 mm long, the tails comprising one-fourth to two-fifths the total length of the seeds; heads 2- to 5- (10-) flowered; stamens 3 or 6_____ _____6. *J. brachycephalus*
            6. Seeds ellipsoid to oblongoid or ovoid, 0.4–0.6 mm long, apiculate.
                8. Stamens 6.
                    9. Involucral leaves shorter than the inflorescences; heads hemispherical or ellipsoid, 2–7 mm wide, 2- to 9-flowered; sepals 1.9–3.0 mm long, acuminate to acute or obtuse; capsules oblongoid or ellipsoid, acute or obtuse_____7. *J. alpinus*
                    9. Involucral leaves usually exceeding the inflorescences; heads hemispherical or spherical, 8–15 mm wide, 9- to 90-flowered; sepals 2.5–5.0 mm long, subulate; capsules lanceoloid, subulate.
                        10. Sepals 2.5–4.0 mm long; petals equalling to exceeding the sepals by 0.8 mm; heads 8–11 (–12) mm wide; stems to 6 dm tall_____ _____8. *J. nodosus*

10. Sepals 4–5 mm long; petals 1 mm shorter than to nearly equalling the sepals; heads 10–15 mm wide; stems to 10.7 dm tall_____9. *J. torreyi*

8. Stamens 3.

    11. Capsules linear-lanceoloid, acute, exceeding the sepals by at least 1.5 mm (usually by 2.0 mm)___ _____10. *J. diffusissimus*

    11. Capsules ellipsoid or oblongoid, obtuse to subulate, shorter than to exceeding the sepals by 1 mm.

        12. Capsules oblongoid, subulate, exceeding the sepals by 0.75–1.00 mm; perianth segments subulate_____11. *J. scirpoides*

        12. Capsules ellipsoid, acute or obtuse, shorter than to exceeding the sepals by 0.75 mm; perianth segments subulate or acuminate.

            13. Perianth segments acuminate; heads 2- to 35-flowered, hemispherical to spherical; petals 0.7 mm shorter than to equalling the sepals; capsules slightly shorter than the petals to exceeding the sepals by 0.75 mm.

                14. Sepals 2.0–2.5 mm long; heads 150–280, 3–5 mm wide, 2- to 7- (8-) flowered; leaves 1.5–4.5 mm wide_____ _____12. *J. nodatus*

                14. Sepals 3–4 mm long; heads (1–) 2–82, 5–10 mm wide, 5- to 35-flowered; leaves 1–3 mm wide_____ _____13. *J. acuminatus*

            13. Perianth segments subulate; heads densely 50- to 80-flowered, spherical; petals distinctly (approximately 1 mm) shorter than the sepals; capsules usually 0.5 mm shorter than the petals_____14. *J. brachycarpus*

3. Flowers borne singly on the inflorescence branches, not in heads, prophyllate.

    15. Annual; auricles absent; sepals 4–7 mm long; inflorescences comprising one-fourth to four-fifths of the total height of the plants_____15. *J. bufonius*

    15. Perennial; auricles present; sepals 2.3–6.0 mm long; inflorescences comprising less than one-half of the total height of the plants.

        16. Sepals obtuse; anthers 1 mm long, three times longer than the filaments_____16. *J. gerardii*

16. Sepals acuminate, subulate, or aristate; anthers shorter than to as long as the filaments.

    17. Leaves terete, at least distally; capsules usually exceeding the perianths; seeds 0.5–1.3 mm long.

        18. Leaves involute near the summit of the sheath, becoming closed and terete above; inflorescences 1.5–6.5 (–8.0) cm long; petals acute or obtuse; capsules exceeding the sepals by (0.75–) 1.0–1.6 mm; seeds 0.5–0.6 mm long, apiculate at both ends_____17. *J. greenei*

        18. Leaves terete throughout; inflorescences (1.0–) 2.0–3.5 cm long; petals acuminate or aristate; capsules slightly shorter than to exceeding the sepals by 1 mm; seeds 1.0–1.3 mm long, caudate at both ends_____18. *J. vaseyi*

    17. Leaves flat or involute; capsules shorter than to exceeding the perianths by 0.1 mm; seeds 0.3–0.5 mm long.

        19. Petals 0.2 mm shorter than to exceeding the sepals by 0.5 mm; tips of the inflorescence branches incurved; leaves to 13 cm long_____ _____19. *J. secundus*

        19. Petals 1 mm shorter than to equalling the sepals; tips of the inflorescence branches not incurved; leaves to 30 cm long.

            20. Auricles friable, not firm or rigid, scarious, hyaline, lanceolate, white or brownish, prolonged 1.0–4.5 (–5.0) mm beyond point of insertion; perianth segments spreading _____20. *J. tenuis*

            20. Auricles firm at apex or rigid, cartilaginous or membranous, occasionally prolonged to 2 mm beyond point of insertion; perianth segments spreading or appressed.

                21. Auricles cartilaginous, opaque, rigid, often slightly flaring, obtuse, yellow or orange-brown, less than 1 mm long and usually 0.75 mm long or less; perianth segments spreading; prophylls obtuse to acute_____21. *J. dudleyi*

                21. Auricles membranous, hyaline, not cartilaginous or rigid, usually firm at apex,

pale or brown, very slightly prolonged
to exserted 2 mm beyond point of insertion; perianth segments appressed;
prophylls acuminate to aristate_____
_____22. *J. interior*

1. **Juncus effusus** L. var. **solutus** Fern. & Wieg. Rhodora 12:
90. 1910. *Fig. 87.*

Juncus bogotensis HBK. var. *solutus* (Fern. & Wieg.) Farw.
Pap. Mich. Acad. I, 23:127. 1937.
Perennial; stems densely cespitose from obscured rhizomes with
inconspicuous internodes, finely many-ribbed, (6–) 10–15 dm
tall, 1.5–4.0 mm wide; basal sheaths 3–4, clasping, brown to
purplish-brown, obtuse, mucronate, the uppermost sheath 9–24
cm long with a mucro 2–4 mm long; involucral leaf erect, terete,
7–48 cm long, exceeding the inflorescence and resembling a
prolongation of the stem; inflorescence appearing lateral on the
stem, irregularly spreading on unequal, compound rays, occasionally rounded and condensed, 1.5–9.0 cm long, 1.75–11.50 cm
wide; perianth segments rigid, lance-attenuate, greenish, brown,
or stramineous; sepals 2.0–3.5 mm long; petals 0.25–0.50 mm,
shorter than to equalling the sepals; stamens 3, anthers 0.5–0.8
mm long, approximately as long as the filaments; capsule brown,
obovoid, truncate or retuse, beakless, 2.5–3.0 mm long, 0.5 mm
shorter than to 0.75 mm longer than the sepals; seeds oblong-
ovoid, 0.5 mm long, minutely apiculate at both ends; 2n = 40
(Darlington & Wylie, 1956).

COMMON NAME: Soft Rush.
HABITAT: Wet ground of shores, swamps, ditches.
RANGE: Newfoundland to North Dakota, south to Texas
and Florida.
ILLINOIS DISTRIBUTION: Throughout the state.
In flowering specimens the petals are 0.25–0.50 mm
shorter than to equalling the sepals and the perianth
exceeds the ovary. During early and late stages of
dehiscence of the capsule, the sepals and/or petals
may be shorter than, equal to, or longer than the capsule. Apparently during the ripening of the ovary the perianth
is spread and the relative lengths of the sepals and petals (with
reference to each other and to the apex of the capsule) become
variable. A specimen from Perry County which represents a

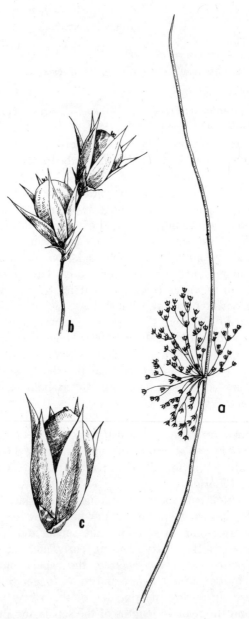

87. *Juncus effusus var. solutus* (Soft Rush). *a*. Habit, X¼. *b*. Portion of ray with capsules, X3½. *c*. Capsule, X6½.

late fruiting stage wherein some of the perianths are below, some equalling, and some slightly exceeding the capsules is known.

According to Gleason (1952), var. *solutus* includes "varieties *pylaei* (LaHarpe) Fern. & Wieg., *decipiens* Buch., and *costulatus* Fern., all of which represent mere phases of fluctuating variability in size and relative length of perianth and capsule."

2. **Juncus balticus** Willd. var. **littoralis** Engelm. Trans. Acad. St. Louis 2:442. 1866. *Fig. 88.*

Juncus balticus var. *littoralis* f. *dissitiflorus* Engelm. ex Fern. & Wieg. Rhodora 25:208. 1923.

Juncus litorum Rydb. Brittonia 1:85. 1931.

Perennial; stems single at intervals from a stout, elongate, occasionally branched rhizome with conspicuous internodes, finely many-ribbed, 3.5–10.5 dm tall, 1.00–2.75 mm wide; basal sheaths purplish or purplish-brown to yellow, often glossy, the uppermost sheath 5–15 cm long, apiculate or with a mucro to 4 mm long; involucral leaf erect, terete, 5–28 cm long, exceeding the inflorescence and resembling a prolongation of the stem; inflorescence appearing lateral on the stem, condensed to diffusely spreading on unequal, compound rays, 1.1–20.0 cm long; perianth segments brown on both sides of a green mid-nerve, the margins scarious and hyaline; sepals lance-attenuate or acuminate, 3.2–5.0 mm long; petals acuminate or acute, 0.5 mm shorter than to nearly equalling the sepals; stamens 6, anthers 1.5- approximately 2.0 mm long, three times longer than the filaments; capsule dark brown, ovoid, acuminate, with a distinct beak 0.5–1.0 mm long, longer than the petals, subequal to 1 mm longer than the sepals; seeds ovoid, 1 mm long, usually minutely apiculate at both ends.

HABITAT: Wet, sandy shores; swamps.

RANGE: Newfoundland to New York and Pennsylvania; about the Great Lakes.

ILLINOIS DISTRIBUTION: Local in the northernmost counties.

In typical var. *littoralis* (Hermann, 1940) the inflorescence is 1.5–3.5 cm long, not diffuse, and the flowers are approximate or subapproximate, whereas f. *dissitiflorus* has a diffuse, remotely flowered inflorescence 4–15 cm long. Fernald & Wiegand (1923) stated that f.

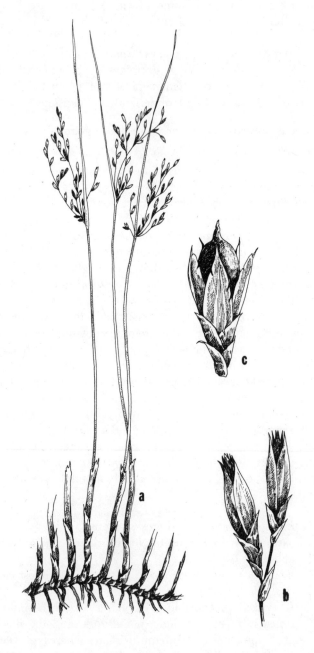

88. *Juncus balticus* var. *littoralis* (Rush). *a*. Habit, X¼. *b*. Portion of ray, X2¾. *c*. Capsule, X3¾.

dissitiflorus "in its most extreme development is very typical of the sand of the Great Lakes, but perfectly characteristic var. *littoralis,* with less diffuse inflorescences and somewhat approximate flowers, also occurs there." Although the majority of Illinois specimens have the long inflorescences of f. *dissitiflorus,* this form is not maintained here because the inflorescences may be condensed or diffuse, with flowers approximate to remote.

3. **Juncus biflorus** Ell. Bot. S. C. and Ga. 1:407. 1817. *Fig. 89.*

Perennial; stems solitary, approximately 3.5 cm apart, from conspicuous, scaly rhizomes; stems 4.9–8.8 (–10.2) dm tall; leaves 2.0–6.5 mm wide; auricles hyaline, pale brown, rounded; involucral leaf inconspicuous, shorter than the inflorescence; inflorescence compact to spreading, the rays ascending to divaricate; inflorescence 2.5–19.5 cm long, heads (13–) 20–135, approximate to distant, 2- to 10-flowered; sepals lanceolate, acuminate or mucronate to short-aristate; petals 2–3 mm long, ovate, obtuse or apiculate, longer than the sepals; stamens 3, nearly equalling the sepals to slightly exceeding the petals, anthers purplish-brown; capsule obovoid, obtuse to truncate, beakless, slightly shorter than to exceeding the petals by 1 mm; seeds oblongoid, approximately 0.5 mm long, apiculate at both ends.

HABITAT: Wet ground of fields, ditches, pastures, prairies.

RANGE: Massachusetts to Michigan, south to Texas and Florida; Mexico.

ILLINOIS DISTRIBUTION: Southern half of the state; Kankakee County. The forms are not distinguished on the map.

Two forms may be separated as follows:

1. Heads predominantly 2- to 4- (5-) flowered_____
_____3a. *J. biflorus* f. *biflorus*
1. Heads predominantly 6- to 10-flowered_____3b. *J. biflorus* f. *adinus*

3a. Juncus biflorus Ell. f. biflorus

Juncus marginatus var. *biflorus* (Ell.) Wood, Class-Book Bot. 725. 1861.

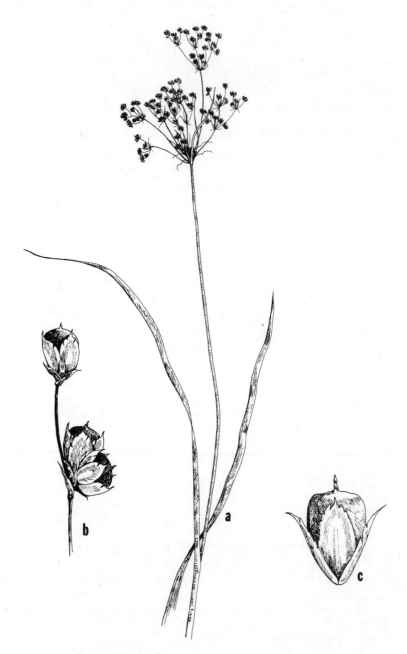

89. *Juncus biflorus* f. *biflorus* (Rush). *a*. Habit, X¾. *b*. Head with capsules, X2½. *c*. Capsule, X3½.

Juncus marginatus var. *aristulatus* (Michx.) Coville, Proc. Biol. Soc. Wash. 8:123. 1893.

3b. **Juncus biflorus** Ell. f. **adinus** Fern. & Grisc. Rhodora 37: 157. 1935.

4. **Juncus marginatus** Rostk. Monog. Junc. 38:pl. 2, f. 3. 1801. *Fig. 90.*

Juncus aristulatus Michx. Fl. Bor. Am. 1:192. 1803.
Tristemon marginatus Raf. Fl. Tellur. 4:32. 1836.
Juncus marginatus var. *vulgaris* Engelm. Trans. Acad. St. Louis 2:455. 1866.
Juncus marginatus var. *paucicapitatus* Engelm. Trans. Acad. St. Louis 2:455. 1866.

Perennial; stems cespitose from a short, scaly, usually inconspicuous rhizome; stems 0.45–5.90 dm tall; leaves 1.0–2.5 (–2.9) mm wide; auricles pale brown; involucral leaf inconspicuous, shorter than the inflorescence; inflorescence compact to spreading, the rays ascending to divaricate; inflorescence (1.5–) 2.5–9.3 (–11.0) cm long, heads 3–28 (–32), approximate to distant, 2- to 10- (15-) flowered; sepals lanceolate, acuminate or mucronate to short-aristate; petals 2.2–3.0 mm long, ovate, obtuse or apiculate, longer than the sepals; stamens 3, nearly equalling the sepals to equalling the petals, anthers purplish-brown; capsule obovoid, obtuse to truncate, beakless, slightly shorter than to equalling the petals; seeds oblongoid, approximately 0.5 mm long, apiculate at both ends; 2n = 38 (Snogerup, 1963).

HABITAT: Wet, often sandy ground of pond borders, ditches, fields.

RANGE: Maine to Ontario, south to Nebraska and Florida.

ILLINOIS DISTRIBUTION: Throughout the state.

The inconspicuous rhizomes of *J. marginatus* occasionally have offset rhizomes to 1.2 cm long. *Juncus biflorus*, in which the culms arise singly approximately 3.5 cm apart, often has offset rhizomes reaching a length of 6.5 cm. Since the flowers of the two species are essentially similar and the numbers of heads and flowers per head are variable, they are usually best separated by observation of the underground parts. Coville (1893) noted that intermediates

90. Juncus marginatus (Rush). *a.* Habit, X¼. *b.* Head with capsules, X2¾. *c.* Capsule, X5.

between *J. marginatus* and *J. biflorus* occur, and considered *J. biflorus* as "probably the mother form" of *J. marginatus*.

5. **Juncus canadensis** J. Gay ex Laharpe, Monogr. Junc. 134. 1825. *Fig. 91.*

Juncus canadensis var. *longecaudatus* Engelm. Trans. Acad. St. Louis 2:474. 1868.
Juncus canadensis var. *paradoxa* Farw. Pap. Mich. Acad. I, 30:60. 1944.
Juncus canadensis var. *typicus* Fern. Rhodora 47:129. 1945.
Juncus canadensis var. *typicus* f. *conglobatus* Fern. Rhodora 47:129. 1945.

Perennial; stems cespitose, 3.0–9.5 dm tall; leaves 3–4, to 3.2 dm long, 1–3 mm wide; involucral leaf shorter than the inflorescence; inflorescence compact to spreading, 2.5–18.0 cm long; heads 5–60 (to approximately 140), approximate to distant, hemispherical to spherical, 5- to 50-flowered; perianth segments lanceolate, acuminate to subulate; sepals 2.3–3.8 mm long; petals mostly exceeding the sepals by up to 0.2 mm, occasionally equalling the sepals; stamens 3, anthers shorter than the filaments; capsule ovoid to lanceoloid, acute or acuminate, slightly shorter than to exceeding the petals by 1.3 mm; seeds fusiform, caudate, 1.2–1.9 mm long, the tail comprising (one-third) one-half to five-eighths of the total length of the seed; 2n = 80 (Snogerup, 1963).

HABITAT: Wet ground of bogs, swamps, meadows.
RANGE: Newfoundland to Minnesota, south to Louisiana and Georgia.
ILLINOIS DISTRIBUTION: Northern half of the state; also Massac, Pope, and Wabash counties.
Evers 32258, from Lake County, is unusual in that the inflorescence contains approximately 140 heads. Farwell's var. *paradoxa* has the inflorescence converted to proliferous heads, as in KANKAKEE CO.: *Hill 293. 1873.*

Certain specimens are referable to f. *conglobatus*. However, this form, wherein the heads are densely crowded into sessile or short-rayed glomerules, seems to be only a phase of the variability in the length of the inflorescence rays in the species. Fernald (1945) stated that specimens of f. *conglobatus* occur which make "every conceivable transition" to typical *J. cana-*

91. *Juncus canadensis* (Rush). *a*. Habit, X¼. *b*. Capsule, X4½. *c*. Seed, X10.

densis which has a more open inflorescence, elongate rays, and mostly scattered heads.

Certain specimens from Kankakee County, *Hill 91.1873* and *Hill 293.1873,* have been indicated by Jones (1945) as the basis for Fernald's inclusion of Illinois in the range of *Juncus brevicaudatus* (Engelm.) Fern.

6. **Juncus brachycephalus** (Engelm.) Buch. Bot. Jahrb. 12: 268. 1890. *Fig. 92.*

Juncus canadensis var. *brachycephalus* Engelm. Trans. Acad. St. Louis 2:474. 1868.

Juncus brachycephalus f. *hexandrus* Martin, Rhodora 40:460. 1938.

Perennial; stems cespitose, 1.4–5.6 dm tall; leaves 3–5, to 12 cm long, 1.0–3.5 mm wide, the auricles membranous; involucral leaf much shorter than the inflorescence; inflorescence generally obpyramidal, ascending or widely spreading, 4–19 cm long; heads 13 to approximately 325, 2–7 mm wide, 2- to 5- (10-) flowered; perianth segments lanceolate, acuminate to obtuse; petals 2.0–2.5 mm long, from scarcely exceeding to exceeding the sepals by up to 0.5 (–0.75) mm; stamens 3 or 6, anthers shorter than the filaments; capsule lanceoloid to narrowly ellipsoid, acuminate, exceeding the petals by up to 1.5 mm; seeds fusiform, 0.8–1.2 mm long, caudate at both ends, the tails comprising one-fourth to two-fifths the total length of the seed.

HABITAT: Marshes, wet ground along bodies of water. RANGE: Nova Scotia to Ontario, south to Illinois and New Jersey.

ILLINOIS DISTRIBUTION: Local in the northern half of the state.

Most inflorescences have three-stamened flowers; rarely an inflorescence will be found with six-stamened flowers. In other inflorescences a head may contain some flowers which have three stamens and others having six.

92. *Juncus brachycephalus* (Rush).   *a.* Habit, X¼.   *b.* Head with capsules, X3½.   *c.* Seed, X12½.

7. **Juncus alpinus** Vill. Histoire des Plantes de Dauphine
2:233. 1787. *Fig. 93.*

Perennial; stems cespitose from a slender rhizome, (5–) 9–30 cm tall; leaves 1–2, to 16 cm long; involucral leaf shorter than the inflorescence; inflorescence branches erect or ascending; inflorescence (2.5–) 3.5–24.0 cm long, heads (2–) 3–38, hemispherical or ellipsoid, 2–7 mm wide, 2- to 9-flowered; flowers sessile or on pedicels to 5 mm long; perianth segments lanceolate; sepals 1.9–3.0 mm long, acuminate to acute or obtuse, usually mucronulate; petals obtuse or acute, occasionally acuminate, rarely apiculate, 0.2–0.5 (–1.0) mm shorter than to rarely equalling the sepals; stamens 6, anthers shorter than the filaments; capsule pale or dark brown, oblongoid or ellipsoid, acute or obtuse, with a mucro to 0.5 mm long, slightly shorter than to exceeding the sepals by 1 mm; seeds ovoid to oblongoid, acute or acuminate, 0.5–0.6 mm long, dark-apiculate at one end; 2n = 40 (Darlington & Wylie, 1956).

HABITAT: Wet sandy shores and marshes.
RANGE: Quebec to British Columbia, south to Washington, Nebraska, Indiana, and Pennsylvania; Eurasia.
ILLINOIS DISTRIBUTION: Local in the northeastern counties. The varieties are not distinguished on the map.
Some Illinois specimens of *J. alpinus* have been confused with *J. articulatus* L. in the past. Fourteen putative specimens examined of *J. articulatus* are *J. alpinus* var. *rariflorus* Hartm. *Juncus articulatus* apparently does not exhibit the key characteristic of *J. alpinus* var. *rariflorus,* that is, having one or several of the flowers of the heads elevated on pedicels above the other flowers of the heads, which often gives the heads an ellipsoid shape. The shapes of the apices of the perianth segments and capsule are variable in *J. alpinus.*

Various authors (Vasey, 1861; Pepoon, 1927; Fernald, 1950; Jones & Fuller, 1955; Winterringer & Evers, 1960; Jones, 1963) have attributed *J. articulatus* to Illinois. The earliest reference by Vasey (1861) is a listing which states *Juncus articulatus* ? L. On the labels of some of Vasey's specimens is written *Juncus articulatus* (KANE CO.: *Vasey s.n.*) or *Juncus*-perhaps new (MCHENRY CO.: *Vasey s.n.*) or *Juncus elongatus? J. articulatus var.* ? (MCHENRY CO.: *Vasey s.n.*) or *Juncus elongatus* ?

93. *Juncus alpinus* var. *rariflorus* (Rush). *a.* Habit, X¼. *b.* Head with capsules, X3. *c.* Seed, X27.

mihi perhaps a new species (MCHENRY CO.: *Vasey s.n.*).
Later authors correctly relegated *J. elongatus* Vasey in herb. to
J. alpinus (Engelmann, 1868; Buchenau, 1906).
Two varieties may be separated by the following key:

1. One or several of the flowers elevated above the others on slightly
elongate pedicels_____7a. *J. alpinus* var. *rariflorus*
1. Flowers sessile or equally short-pedicellate_____
_____7b. *J. alpinus* var. *fuscescens*

7a. Juncus alpinus Vill. var. **rariflorus** Hartm. Skand. Fl., ed.
7:240. 1858.
Juncus richardsonianus Schultes in Roemer & Schultes, Linn.
Syst. Veg. 7:201. 1829.
Juncus alpinus var. *insignis* Fries ex Engelm. Trans. Acad.
St. Louis 2:459. 1868.

7b. Juncus alpinus Vill. var. **fuscescens** Fern. Rhodora 10:
48. 1908.

8. Juncus nodosus L. Sp. Pl. ed. 2:466. 1762. *Fig. 94.*
Juncus nodosus var. *vulgaris* Torr. Fl. N. Y. 2:326. 1843.
Juncus nodosus var. *genuinus* Engelm. Trans. Acad. St. Louis
2:471. 1868.
Juncus nodosus var. *proliferus* Lunell, Am. Midl. Nat. 4:238.
1915.
Perennial from rhizomes with numerous tuberous thickenings;
stems to 6 dm tall, to 3 mm wide; leaves 2–5, erect or ascending, 1–2 dm long, 1.0–1.5 (–2.0) mm wide; auricles membranous, yellowish, prolonged 0.5–1.0 (–3.0) mm beyond point of
insertion; involucral leaf exceeding the inflorescence; inflorescence compact to spreading, 1–5 (–7) cm long; heads (1–) 2–
15, hemispherical or spherical, 8–11 (–12) mm wide, 9- to 35-
flowered; perianth segments lanceolate, subulate; sepals 2.5–
4.0 mm long; petals equalling to exceeding the sepals by 0.8 mm;
stamens 6, anthers shorter than the filaments; capsule lanceoloid, subulate, equalling to exceeding the perianth by 1.5 mm;
seeds ovoid or oblongoid, 0.5 mm long, apiculate at both ends.

94. *Juncus nodosus* (Rush). *a*. Habit, X¼. *b*. Flowering heads, X1⅝. *c*. Capsule, X5½.

HABITAT: Wet, often sandy ground of marshes, swamps, fields, shores.

RANGE: Newfoundland to Alaska, south to New Mexico, Illinois, and Virginia.

ILLINOIS DISTRIBUTION: Northern third of the state; Washington County.

The untenable var. *proliferus*, wherein the flowers are replaced by tufts of leaves, is known from Cook and McHenry counties.

9. **Juncus torreyi** Coville, Bull. Torrey Club 22:303. 1895. *Fig. 95.*

Juncus nodosus L. var. *megacephalus* Torr. Fl. N. Y. 2:326. 1843.

Juncus megacephalus Wood, Class-Book Bot. ed. 2, 724. 1861, non Curtis (1835).

Juncus torreyi var. *proliferus* Lunell, Am. Midl. Nat. 4:239. 1915.

Juncus torreyi f. *longipes* Farw. Pap. Mich. Acad. I, 1:91. 1921.

Juncus torreyi f. *brepipes* Farw. Pap. Mich. Acad. I, 1:92. 1921.

Juncus torreyi var. *paniculata* Farw. Pap. Mich. Acad. I, 1:92. 1921.

Juncus torreyi var. *globularis* Farw. Pap. Mich. Acad. I, 1:92. 1921.

Perennial from rhizomes with numerous tuberous thickenings; stems to 10.7 dm tall, to 5 mm wide; leaves 2–5, often divaricate, 1.0–4.9 dm long, 1–3 mm wide; auricles membranous and yellowish or hyaline, prolonged 2.5–4.0 mm beyond point of insertion; involucral leaf often exceeding the inflorescence; inflorescence compact to spreading, 1–15 cm long; heads 1–21 (–45), spherical, 10–15 mm wide, (14-) 25- approximately 90-flowered; perianth segments lanceolate, subulate; sepals 4–5 mm long; petals 1 mm shorter than to nearly equalling the sepals; stamens 6, anthers shorter than the filaments; capsule lance-subuloid, slightly shorter than to exceeding the sepals by 1 mm; seeds ovoid or oblongoid, 0.5 mm long, minutely apiculate at both ends; 2n = 40 (Snogerup, 1963).

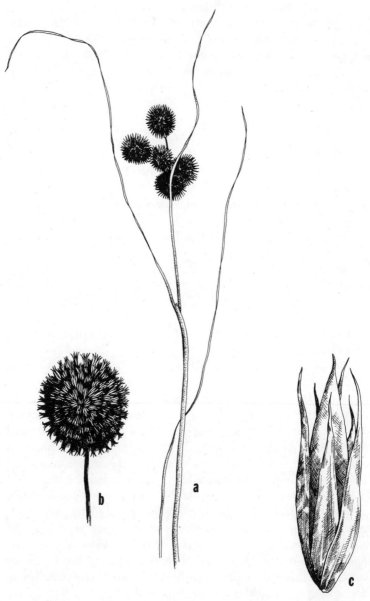

95. *Juncus torreyi* (Rush). *a.* Habit, X¼. *b.* Flowering head, X1⅛. *c.* Capsule, X6¾.

HABITAT: Wet ground.

RANGE: Massachusetts to Saskatchewan and Washington, south to Arizona, Texas, and Alabama.

ILLINOIS DISTRIBUTION: Throughout the state.

In some plants the flowers are replaced by tufts of leaves. From the same rhizome there may be stems which have the inflorescence wholly converted to leaves, stems with unchanged flowers, and stems, the inflorescences of which have heads partially leafy and partially composed of flowers. Examples of this untenable variety *proliferus* are known from Cook, Peoria, and Winnebago counties.

In 1921, Farwell named two new forms and two new varieties of *Juncus torreyi*. He stated that "the typical form of this species has few (usually 3–6) very large, pale-brownish heads (50- to 80-flowered, 14–18 mm in diameter) composing a short (2–3 cm), simple panicle, terminating a stout stem often 1 m in height." His "typical form" corresponds to specimens seen from Will, Cook, and Henry counties. Farwell stated that from the typical form there are "several distinct forms varying to *J. nodosus* L. in the size of heads and slenderness of stems, but retaining the floral characters of *J. torreyi* Coville." Illinois specimens have been examined which are referable to Farwell's forms *longipes* and *brepipes* (*sic*) and varieties *globularis* and *paniculata*.

Forma *longipes*, which differs from the typical form in having an inflorescence 4–7 cm long on long peduncles with 15–30 heads, 10–12 mm wide has been collected in DuPage County. However, numerous specimens with long peduncles have measurements which hover around those given to f. *longipes* by Farwell.

Forma *brepipes* is similar to f. *longipes*, but the peduncles are so abbreviated that the glomerules form a dense, often irregular head. An example of this form, TAZEWELL CO.: July, 1891, *McDonald s.n.*, has a congested ovoid inflorescence 6 cm long with about 21 approximate heads.

Variety *globularis*, in which the stem is terminated by a single globular head of 40–80 flowers, 15 mm wide, is known from Cook, McHenry, and Perry counties. Farwell designated the stem height of var. *globularis* as about 3 dm; in the above mentioned collections it varies from 3.6–7.7 dm.

Concerning var. *paniculata*, Farwell stated "Panicle oblong

(16 cm) with 20–60 or more pale-brown heads (10- to 30-flowered, 8–10 mm in diameter). In the small, few-flowered heads, this variety approaches and connects with *J. nodosus;* some of the flowers have the petals equal in length to the sepals as in *J. nodosus,* but lack the long capsule and the brownish-purple color characteristic of that species." Specimens referable to var. *paniculata* are: MCHENRY CO.: Ringwood, *Vasey s.n.,* inflorescence 9 cm long, 45 heads, 14 to approximately 25 flowers per head, heads 9–10 mm wide; Ringwood, *Vasey s.n.,* inflorescence 10–14 cm long, approximately 38 heads. The first mentioned Vasey specimen has a label with the notation "Juncus nodosus forma intermedia major paniculata"; the petals are shorter than to nearly equalling the sepals. The other Vasey specimen has the petals equalling the sepals in some flowers and petals shorter than the sepals in other flowers.

Coville (1895) stated that *J. torreyi* and *J. nodosus* do not intergrade, whereas Engelmann (1868) mentioned intermediate Vasey specimens.

Farwell's forms *longipes* and *brepipes,* as well as var. *globularis,* appear to be merely growth phases of *J. torreyi* and their status is not retained here. *Juncus torreyi* is undoubtedly very closely related to *J. nodosus,* and although intermediates between the two species occur (referable to *J. torreyi* var. *paniculata*), it is deemed best to consider *J. nodosus* and *J. torreyi* separate species.

10. **Juncus diffusissimus** Buckl. Proc. Acad. Phila. 14:9. 1862.
   Fig. 96.

Juncus acuminatus Michx. var. *diffusissimus* (Buckl.) Engelm. Trans. Acad. St. Louis 2:463. 1868.

Perennial; stems cespitose, (0.9–) 1.1–2.6 dm tall, 1–2 mm wide; leaves 2–3, 0.5–2.0 mm wide; involucral leaf shorter than the inflorescence; inflorescence diffusely branched, spreading, (6–) 8–16 cm long, comprising two-fifths to one-third the height of the plant in robust individuals; heads approximately 50–150, as few as 4–9 in small specimens, hemispherical or obpyramidal, 2- to 10-flowered; perianth segments lanceolate, subulate; sepals 2.2–3.0 mm long; petals equalling to often slightly shorter than the sepals; stamens 3, anthers shorter than the filaments; capsule linear-lanceloid, acute, mucronulate, exceeding the perianth by at least 1.5 mm and usually 2 mm, rarely more; seeds narrowly ovoid, 0.4–0.5 mm long.

96. *Juncus diffusissimus* (Rush). *a*. Habit, X¼. *b*. Head with capsules, X2½. *c*. Capsule, X5¾.

HABITAT: Wet ground, ditches.

RANGE: New York to Kansas, south to Texas and Georgia; Virginia to South Carolina.

ILLINOIS DISTRIBUTION: Local in the southern third of the state.

The first Illinois collection of this species is *J. P. Sivert s.n.* from Lawrence County in 1952.

11. Juncus scirpoides Lam. Encycl. Meth. Bot. 3:267. 1789. *Fig. 97.*

Juncus scirpoides var. *genuinus* Buch. Bot. Jahrb. 12:323. 1890.

Perennial from short rhizome; stems cespitose, (0.5–) 1.0–4.6 dm tall, 0.50–2.75 mm wide; leaves 2–3, 1–2 mm wide, to 2.2 dm long; auricles membranous, pale, oblong, acute, prolonged 1–2 mm beyond point of insertion; involucral leaf shorter than to slightly exceeding the inflorescence; inflorescence compact to spreading with several rays, (0.5–) 2–15 cm long; heads 1–34, spherical, 7–12 mm wide, 20- to approximately 70-flowered; perianth segments lanceolate, subulate; sepals (2.5–) 3.0–3.5 mm long; petals nearly equalling to approximately 1 mm shorter than the sepals; stamens 3, anthers shorter than the filaments; capsule oblongoid, subulate, exceeding the sepals by 0.75–1.00 mm; seeds ovoid, 0.4–0.5 mm long, apiculate at both ends.

HABITAT: Wet and often sandy ground.

RANGE: New York to Michigan, south to Texas, Missouri, and Florida.

ILLINOIS DISTRIBUTION: Cass, Lawrence, and Menard counties.

12. Juncus nodatus Coville in Britt. & Brown, Illustr. Fl. N. U. S., ed. 2, 1:482. 1913. *Fig. 98.*

Juncus acuminatus Michx. var. *robustus* Engelm. Trans. Acad. St. Louis 2:463. 1868.

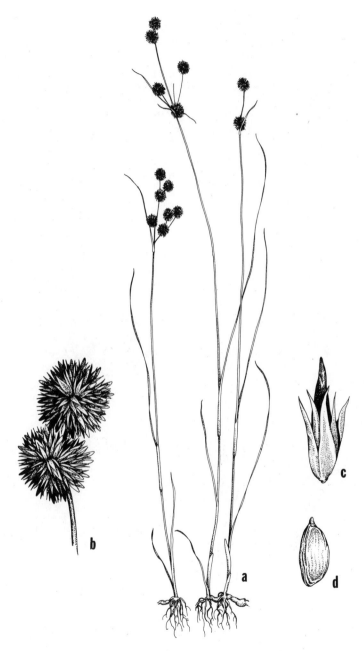

97. Juncus scirpoides (Rush). *a.* Habit, X¼. *b.* Heads with capsules, X1.
c. Capsule, X4. *d.* Seed, X18½

98. Juncus nodatus (Rush). *a.* Habit, X$\frac{1}{12}$. *b.* Heads with capsules, X3½.
c. Capsule, X5½.

Juncus robustus (Engelm.) Coville in Britt. & Brown, Illustr. Fl. N. U. S., ed. 2, 1:395. 1896, non Wats. (1879).

Perennial; stems single or two together, 5.9–8.3 dm tall, 3–10 mm wide; leaves strongly nodose-septate, 1.5–4.5 mm wide, to 5.1 dm long; involucral leaf much shorter than the inflorescence; inflorescence diffuse with numerous, widely divergent rays, 1.5–2.7 dm long; heads 150–280, 3–5 mm wide, hemispherical, 2- to 7- (8-) flowered; perianth segments lanceolate, acuminate; sepals 2.0–2.5 mm long; petals slightly shorter than to equalling the sepals; stamens 3; capsule ellipsoid, acute or obtuse, blunt or mucronulate, nearly equalling to exceeding the sepals by 0.75 mm; seeds narrowly ellipsoid, 0.5–0.6 mm long, minutely apiculate at both ends.

HABITAT: Wet grounds, ditches, edges of water bodies. RANGE: Delaware to Kansas, south to Texas and Florida.

ILLINOIS DISTRIBUTION: Southern half of the state; also Hancock County.

Some specimens of *Juncus nodatus* have fewer heads than is typical for the species. Examples of this condition are known from Cumberland, Pope, and Williamson counties.

Other atypical specimens have all the features of the species, but have 10–12 flowers per head. The length of the perianth (2.0–2.5 mm) is probably the most constant diagnostic characteristic of *J. nodatus* which separates it from *J. acuminatus*.

13. **Juncus acuminatus** Michx. Fl. Bor. Am. 1:192. 1803.

Fig. 99.

Juncus paradoxus E. Mey. Syn. Junc. 30. 1822.

Juncus acuminatus var. *legitimus* Engelm. Trans. Acad. St. Louis 2:463. 1868.

Juncus acuminatus var. *paradoxus* (E. Mey.) Farw. Am. Midl. Nat. 11:73. 1928.

Juncus acuminatus f. *sphaerocephalus* Hermann, Leaflets Western Botany 8:13. 1956.

Perennial from a short, inconspicuous rhizome; stems cespitose, 1.3–7.5 dm tall, 1–3 (–4) mm wide; leaves 1–3 mm wide, to 3.4 dm long; involucral leaf shorter than the inflorescence; inflorescence compact to spreading, (0.5–) 2–14 (–18) cm long; in-

99. *Juncus acuminatus* (Rush). *a.* Habit, X¼. *b.* Heads, X1½. *c.* Capsule, X6½.

florescence rays ascending or divergent; heads (1-) 2–82, hemispherical to spherical, 5–10 mm wide, 5- to approximately 35-flowered; perianth segments lanceolate, acuminate; sepals 3–4 mm long; petals 0.7 mm shorter than to equalling the sepals; stamens 3, anthers shorter than the filaments; capsule ellipsoid, acute, mucronate, slightly shorter than the petals to exceeding the sepals by 0.3 mm; seeds ellipsoid, 0.5 mm long, minutely apiculate at both ends.

HABITAT: Wet, low ground of ditches; ponds.
RANGE: Maine and Nova Scotia to Minnesota, south to Texas and Florida; British Columbia to Oregon; Mexico.
ILLINOIS DISTRIBUTION: Throughout the state.
A specimen from Williamson County, *Bell s.n.*, is uncharacteristic. It has, in the majority of the inflorescences, heads of an ovoid to ellipsoid shape. Rather than being clustered at one point in hemispherical or spherical heads as is typical in the species, the flowers are clustered along slightly elongated extensions of the peduncles.

14. **Juncus brachycarpus** Engelm. Trans. Acad. St. Louis 2:467. 1868. *Fig. 100.*

Perennial from a rhizome 3–4 mm wide; stems one or several, (1.5-) 2.3–9.8 dm tall; leaves (2-) 3 (–5), erect or slightly spreading, 1–2 (–3) mm wide; auricle scarious, acute or truncate, prolonged 2–3 mm beyond point of insertion; involucral leaf shorter than to exceeding the inflorescence by 2 cm; inflorescence compact to open and branched, (1-) 2–9 (–11) cm long; heads 3–10 (–15), spherical, approximately 1 cm wide, densely 50- to 80-flowered; perianth segments lanceolate, subulate; sepals 3–4 mm long; petals 2–3 mm long, distinctly shorter than the sepals; stamens 3, anthers shorter than the filaments; capsule drab brown, ellipsoid, acute, mucronate, usually shorter than the petals by 0.5 mm, rarely equalling the petals; seeds 0.4 mm long, apiculate at both ends.

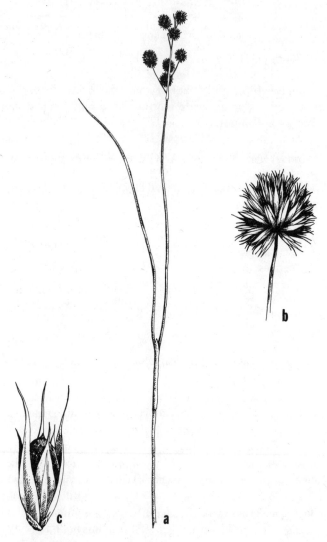

100. Juncus brachycarpus (Rush). *a.* Habit, X¼. *b.* Flowering head, X1⅛. *c.* Capsule, X5¾.

HABITAT: Low, wet ground of fields, prairies, ditches, roadsides.

RANGE: Massachusetts to Ontario, south to Texas, Mississippi, and Georgia.

ILLINOIS DISTRIBUTION: Throughout the state.

In this species the involucral leaf usually does not exceed the inflorescence; in *Evers 23859*, from Fayette County, the involucral leaf extends 2 cm beyond the inflorescence. The capsules of *J. brachycarpus* are usually shorter than the petals, often by 0.5 mm. Only rarely will a specimen be found which has flowers whose capsules equal the petals. A specimen from Pulaski County, *Winterringer 11128*, is atypical with heads 6–7 mm in diameter and 21 to 45 heads per stem.

15. Juncus bufonius L. Sp. Pl. 328. 1753. *Fig. 101.*

Annual with fibrous roots; stems single, often branching at base of plant and simulating a crown of cespitose stems, 1.0–12.5 cm tall below the inflorescence, 0.3 to approximately 1.0 mm wide; leaves 1–4, flat or involute, 0.3–1.0 mm wide, auricles absent; involucral leaf shorter than the inflorescence; inflorescence spreading with forked branches, comprising one-fourth to four-fifths of the total height of the plant; flowers remote or in 2- to 4-flowered fascicles, mostly secund, subsessile, or on pedicels to 1 mm long, prophylls acute or obtuse; perianth segments lanceolate, long-acuminate; sepals 4–7 mm long; petals 3.5–5.0 mm long; stamens 3 or 6, anthers 0.5–1.0 mm long; capsule obovoid, obtuse or truncate, mucronate, beakless, 2.5–4.0 mm long, 0.5–1.0 mm shorter than the petals, green to red-brown, trilocular; seeds oblongoid to ovoid, 0.5 mm long, minutely apiculate at one or both ends; 2n = *ca.* 60, *ca.* 120 (Darlington & Wylie, 1956).

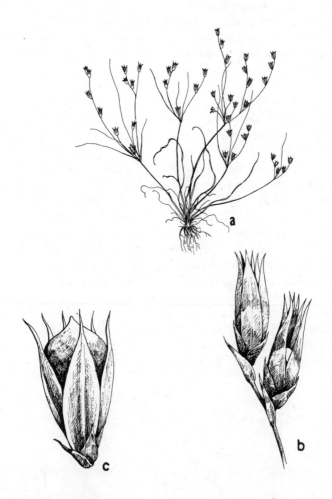

101. *Juncus bufonius* var. *bufonius* (Toad Rush). *a.* Habit, X¼. *b.* Ray, X2½. *c.* Capsule, X7½.

COMMON NAME: Toad Rush.

HABITAT: Wet roadsides, fields.

RANGE: Almost throughout North America; cosmopolitan.

ILLINOIS DISTRIBUTION: Local throughout the state. The varieties are not distinguished on the map. Plants of *J. bufonius*, the only annual species in Illinois, attain a height of 4–17 cm. The green or brown midnerves of the sepals extend to the apex. The midnerves of the petals do not extend to the apex, and their apices are thus translucent. Cleistogamous flowers have been observed in this species (DeFilipps, 1964).

Two varieties have been collected in Illinois:

1. Terminal flowers and flowers on the inflorescence branches borne singly, separated_____15a. *J. bufonius* var. *bufonius*
1. Many of the terminal flowers and flowers on the inflorescence branches closely aggregated into 2- and 4-flowered fascicles_____
_____15b. *J. bufonius* var. *congestus*

15a. Juncus bufonius L. var. bufonius

Juncus bufonius var. *genuinus* Coutinho, Bol. Soc. Brot. 8:102. 1890.

This is the more common variety.

15b. Juncus bufonius L. var. congestus Wahlb. Flora Gothoburgensis 38. 1820.

16. Juncus gerardii Loisel. Jour. Bot. Desvaux 2:284. 1809.

Fig. 102.

Juncus floridanus Raf. Aut. Bot. 194. 1840.

Perennial; stems loosely cespitose from a conspicuous rhizome, 2.0–5.7 dm tall; leaves 1–2 (–3), involute, 15–26 cm long, 0.5–1.0 mm wide, auricles membranous, pale, not prolonged beyond point of insertion; involucral leaf shorter than to exceeding the inflorescence by 5 mm; inflorescence narrow, 5–10 cm long, the branches erect or ascending, prophylls acute; perianth segments oblong, obtuse; sepals 2.3–3.0 mm long; petals subequal to equalling the sepals; stamens 6, anthers 1 mm long, three times longer than the filaments; capsule ovoid, obtuse, mucronate, equalling to exceeding the sepals by 0.8 mm; seeds ovoid, apiculate, 0.4–0.5 mm long; $2n = 80, 84$ (Darlington & Wylie, 1956; Snogerup, 1963).

102. Juncus gerardii (Rush). *a.* Habit, X¼. *b.* Portion of ray, X3. *c.* Seed, X19½

COMMON NAME: Black Grass.

HABITAT: Disturbed marshy area.

RANGE: Quebec to Florida; British Columbia to Washington; Eurasia; North Africa.

ILLINOIS DISTRIBUTION: Adventive in Chicago, Cook County; collected by S. F. Glassman in 1956.

17. **Juncus greenei** Oakes & Tuckerm. in Tuckerm. Am. Jour. Sci. 45:37. 1843. *Fig. 103.*

Perennial; stems cespitose from an inconspicuous rhizome, 3.0–7.4 dm tall, erect; leaves involute near the summit of the sheath, becoming closed and terete above, erect or flexuous, to 3.3 dm long; auricles membranous or rigid, short, obtuse or acute, yellowish or pale brown to black, not conspicuously prolonged beyond point of insertion; involucral leaves 1 or 2, the longest exceeding the inflorescence by up to 18 cm; inflorescence compact to rarely slightly open, 1.5–6.5 (–8.0) cm long, few- to many-flowered; flowers approximate, prophylls short, obtuse; perianth segments lanceolate or lance-ovate, appressed; sepals acuminate, aristulate, 2.5–4.0 mm long; petals acute or obtuse, mucronulate or emucronulate, shorter than the sepals by up to approximately 0.75 mm, less commonly equalling the sepals; stamens 6, anthers oblong-linear, as long as the filaments; capsule yellowish- to reddish-brown, ovoid or oblongoid, usually truncate, less commonly obtuse, retuse, longer than the petals and exceeding the sepals by (0.75–) 1–1.6 mm; placentation axile; seeds oblongoid or ellipsoid, 0.5–0.6 mm long, apiculate at both ends.

HABITAT: Wet, usually sandy ground of meadows, prairies, and fields.

RANGE: Maine to Vermont and New Jersey; inland in Indiana, Illinois, Iowa, and Minnesota.

ILLINOIS DISTRIBUTION: Local in the northeastern counties.

103. Juncus greenei (Rush).   *a.* Habit, X1/10.   *b.* Ray with capsules, X2¼.
c. Seed, X25¾.

18. **Juncus vaseyi** Engelm. Trans. Acad. St. Louis 2:448. 1866.
Fig. 104.

Perennial; stems cespitose from an inconspicuous rhizome, 1 mm wide, 3.7–6.9 dm tall, erect; leaves 1 or 2, terete throughout entire length of blade, 2.1–4.0 dm long, 0.8 mm wide; auricles semimembranous, rounded, yellowish or brown, not prolonged beyond point of insertion; involucral leaf shorter than to exceeding the inflorescence by 4 cm; inflorescence compact with erect or ascending rays, (1.0–) 2.0–3.5 cm long, few-flowered; flowers mostly approximate, prophylls obtuse or acute; perianth segments lanceolate, acuminate or aristate, appressed; sepals 3.3–4.6 mm long; petals 0.5 mm shorter than to equalling the sepals; stamens 6, anthers as long as the filaments; capsule greenish- or pale yellowish-brown, ellipsoid-oblongoid, obtuse or truncate, retuse, slightly shorter than to exceeding the sepals by 1 mm; placentation axile; seeds fusiform, unequally caudate, 1.0–1.3 mm long, the bodies 0.5–0.8 mm long, the tails 0.2–0.4 mm long.

HABITAT: Wet meadows, bogs, shores.
RANGE: Quebec to Alberta, south to Utah, Colorado, Illinois, Michigan, New York, and Maine.
ILLINOIS DISTRIBUTION: Local in the northeastern counties.
The type specimen, *Vasey s.n.*, from McHenry County, is in the Missouri Botanical Garden herbarium.

19. **Juncus secundus** Beauv. in Poiret, Encycl. Method. Bot. Suppl. 3:160. 1813. *Fig. 105.*

Juncus tenuis var. *secundus* Engelm. Trans. Acad. St. Louis 2:450. 1866.

Perennial; stems cespitose, 1.5–3.7 dm tall; leaves 1–2, flat, to 13 cm long, 0.5–1.0 mm wide, auricles membranous, stramineous or dark at apex, not prolonged beyond point of insertion; involucral leaves 2, unequal in length, shorter than the inflorescence, rarely exceeding the inflorescence by 6 mm; inflorescence 3.0–7.5 cm long, inflorescence branches erect to slightly spreading, usually ascending, the tips of the branches incurved; flow-

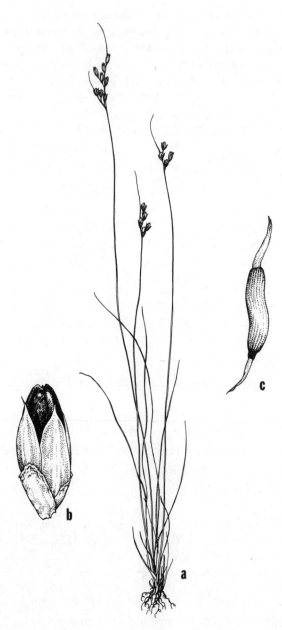

104. Juncus vaseyi (Rush).   *a.* Habit, X⅛.   *b.* Capsule, X3¾.   *c.* Seed, X22.

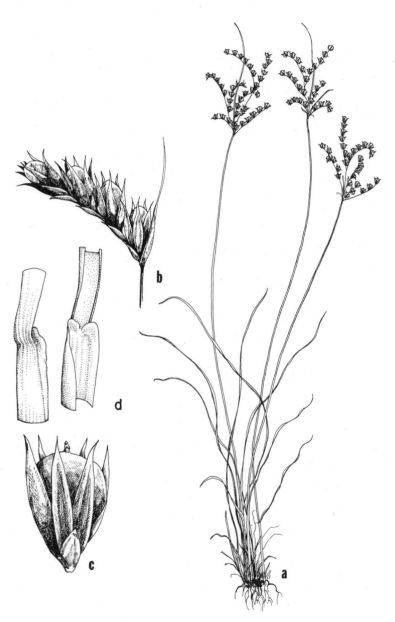

105. *Juncus secundus* (Rush). *a.* Habit, X¼. *b.* Ray with capsules, X2½. *c.* Capsule, X5¾. *d.* Sheaths, showing auricles, X3½.

ers secund, approximate or occasionally distant, prophylls acute
or acuminate; perianth segments lanceolate, acuminate or subulate, erect to slightly spreading; sepals 2.5–3.6 mm long; petals
0.2 mm shorter than to exceeding the sepals by 0.5 mm; stamens
6, anthers approximately as long as the filaments; capsule trilocular, ovoid, obtuse or truncate, retuse, apiculate, subequal
to exceeding the perianth by 0.1 mm; seeds oblongoid or ovoid,
0.3–0.4 mm long, apiculate at both ends.

HABITAT: Open sandstone ledges and outcroppings.
RANGE: Maine to Vermont and North Carolina; Ten-
nessee to Indiana, Illinois, and Missouri.
ILLINOIS DISTRIBUTION: Apparently confined to the
Shawnee Hills region of southern Illinois.

20. Juncus tenuis Willd. Sp. Pl. 2:214. 1799. *Fig. 106.*

Juncus bicornis Michx. Fl. Bor. Am. 1:191. 1803.

Juncus macer S. F. Gray, Nat. Arr. Brit. Pl. 2:164. 1821.

Juncus tenuis var. *anthelatus* Wieg. Bull. Torrey Club
27:523. 1900.

Juncus tenuis var. *williamsii* Fern. Rhodora 3:60. 1901.

Juncus bicornis var. *williamsii* (Fern.) Vict. Contrib. Lab.
Bot. Univ. Montreal 14:32. 1929.

Juncus macer var. *anthelatus* (Wieg.) Fern. Journ. Bot.
68:367. 1930.

Juncus macer var. *williamsii* (Fern.) Fern. Journ. Bot.
68.367. 1930.

Juncus macer f. *anthelatus* (Wieg.) Hermann, Rhodora
40:81. 1938.

Juncus macer f. *discretiflorus* Hermann, Rhodora 40:82. 1938.

Juncus macer f. *williamsii* (Fern.) Hermann, Rhodora 40:82.
1938.

Juncus tenuis f. *anthelatus* (Wieg.) Hermann, Castanea
10:23. 1945.

Juncus tenuis f. *discretiflorus* (Hermann) Fern. Rhodora
47:123. 1945.

Juncus tenuis f. *williamsii* (Fern.) Hermann, Castanea 10:23.
1945.

106. Juncus tenuis (Path Rush). *a.* Habit, X¼. *b.* Ray with capsules, X2. *c.* Capsule, X4. *d.* Sheaths, showing auricles, X3½.

Perennial; stems cespitose, 1–6 dm tall, erect or spreading; leaves flat or involute, 0.5–1.5 mm wide; auricles scarious, hyaline, friable, lanceolate, white or brownish, prolonged 1.0–4.5 (–5.0) mm beyond point of insertion; involucral leaves 2 or 3, the longest usually exceeding the inflorescence by up to 12 cm; inflorescence compact to open, the rays erect to spreading; inflorescence 0.1–1.9 dm long, few- to many-flowered; flowers secund and approximate to widely scattered, prophylls obtuse or acute; perianth segments lanceolate, acuminate or subulate, spreading; sepals 3–5 mm long; petals 0.5 mm shorter than to equalling the sepals; stamens 6, anthers shorter than the filaments; capsule ovoid or oblongoid-ovoid, obtuse or retuse, apiculate, usually one-half to three-fourths the length of the sepals, occasionally equalling the sepals; seeds oblongoid or ovoid, 0.3–0.4 mm long, apiculate at both ends; 2n = 30, 32 (Darlington & Wylie, 1956).

COMMON NAME: Path Rush.

HABITAT: Moist or dry ground.

RANGE: Almost throughout North America; adventive in Europe, South America, and Australia.

ILLINOIS DISTRIBUTION: Very common; throughout the state.

Juncus tenuis has several named forms which are distinct in their extreme types. The forms are probably ecological phases. However, numerous specimens transitioning from f. *tenuis* to other forms are found.

In f. *tenuis*, flowers are mostly clustered at the tips of the inflorescence branches; in f. *williamsii*, the flowers are scattered or secund on widely spreading branchlets to 1 (–2) centimeters long; in f. *anthelatus*, flowers are scattered or secund on ascending branchlets to 4 centimeters long; and in f. *discretiflorus*, the scattered or secund flowers are on ascending branchlets mostly longer than 4 centimeters.

21. Juncus dudleyi Wieg. Bull. Torrey Club 27:524. 1900.
Fig. 107.

Juncus tenuis var. *dudleyi* (Wieg.) Hermann in Johnston, Journ. Arnold Arb. 25:56. 1944.

Perennial; stems cespitose, 3–9 dm tall; leaves flat or involute, 0.5–1.0 mm wide, 1–3 dm long; auricles cartilaginous, opaque,

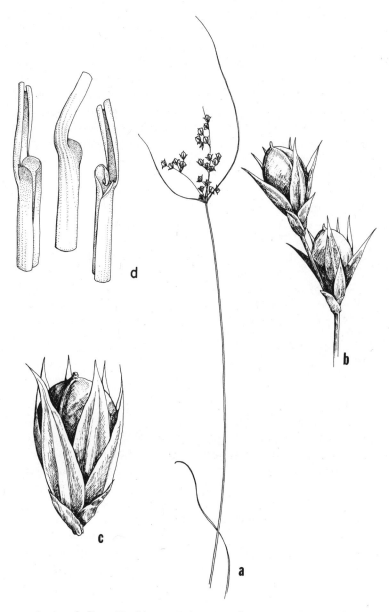

107. *Juncus dudleyi* (Rush). *a.* Habit, X¼. *b.* Portion of ray with capsules, X2. *c.* Capsule, X4¼. *d.* Sheaths, showing auricles (left auricle
cut away in illustration on extreme right), X3½.

rigid, often slightly flaring, rounded at apex, yellow, yellow-
to orange-brown, less than 1 mm long and usually 0.75 mm
long or less, not conspicuously prolonged beyond point of insertion; involucral leaves usually 2, exceeding the inflorescence by
up to 13 cm; inflorescence rays erect to spreading; inflorescence
1–7 (–8) cm long; flowers approximate to distant, occasionally
secund on spreading rays; prophylls ovate, obtuse to acute;
perianth segments lanceolate, acuminate, spreading; sepals
(3.5–) 4.0–6.0 mm long, petals slightly shorter than to equalling
the sepals; stamens 6, anthers shorter than the filaments; capsule
ovoid, obtuse or truncate, apiculate, usually three-fourths to
seven-eighths the length of the sepals; seeds oblongoid or
ovoid, 0.4–0.5 mm long, apiculate at both ends.

HABITAT: Moist ground.
RANGE: Newfoundland to Washington, south to Arizona, Texas, Tennessee, and Maryland; Mexico.
ILLINOIS DISTRIBUTION: Throughout the state.
Some specimens show a light red color at the point
of insertion of the auricles. The auricles are cartilaginous, and the perianths are characteristic of *J. dudleyi.*
Certain specimens of *J. interior* have light red-colored
auricles also.

22. Juncus interior Wieg. Bull. Torrey Club 27:516. 1900.
Fig. 108.

Perennial; stems cespitose, (1–) 3–10 dm tall; leaves flat or involute, 0.5–2.0 mm wide, to 25 cm long; auricles membranous,
hyaline, usually firm at apex, not rigid, pale or stramineous,
often tinged with brown at apex or point of insertion, usually
not prolonged beyond point of insertion, involucral leaves usually 2, shorter than to exceeding the inflorescence by 13 cm;
inflorescence compact with erect rays to spreading, 1–12 cm
long; flowers approximate to distant, occasionally secund; prophylls acuminate to aristate; perianth usually appressed to capsule, occasionally slightly spreading, perianth segments lanceolate, acuminate, subulate, or aristulate; sepals 3.5–6.0 mm long;
petals 1 mm shorter than to equalling the sepals; stamens 6,
anthers shorter than the filaments; capsule ovoid, obtuse or
truncate, apiculate, from three-fourths the length to equalling
the sepals; seeds oblongoid or ovoid, 0.4–0.5 mm long, apiculate
at both ends.

108. Juncus interior (Rush). *a.* Habit, X⅕. *b.* Portion of ray with capsules, X1½. *c.* Capsule, X4¼. *d.* Sheaths, showing auricles, X3½.

HABITAT: Wet ground of fields, prairies, ditches, road-sides.

RANGE: Ohio to Michigan, south to Arizona, Missouri, and Indiana.

ILLINOIS DISTRIBUTION: Throughout the state.

In most specimens the longest involucral leaf exceeds the inflorescence. The auricles are thin and membranous and, although usually firm, are not rigid at the apex in comparison to the auricles of *J. dudleyi*. There is often a tinge of brown color on the apex of the auricles or at the point of their insertion on the sheath; less frequently the auricles are tinged with light red. Usually the auricles are either gradually curved and only very slightly prolonged or they are rounded at the apex and exserted to 1 mm beyond the point of insertion. However, some specimens of *J. interior* which possess the characteristic appressed perianths and acuminate prophylls of the species have auricles prolonged 1.5–2.0 mm. The type specimen, collected by G. Vasey in McHenry County, is in the Gray Herbarium.

Species Excluded

Calla palustris L. Brendel first reported the water arum from Illinois in 1859, and McDougall, in 1936, reported that the species occurred in cold bogs in Lake and McHenry counties. I have been unable to find any specimens of this species preserved from Illinois. The natural range is slightly to the north and east of Illinois.

Eriocaulon septangulare With. Although the pipewort is found in Wisconsin and Indiana, there are no specimens known from Illinois despite Muenscher's report of it from this state in 1944.

Juncus brevicaudatus (Engelm.) Fern. Despite Fernald's report of this species from Illinois in 1950, the specimen on which his report is based is *J. canadensis* J. Gay ex Laharpe.

Orontium aquaticum L. The golden club, an aquatic eastern perennial known as far west as West Virginia and Kentucky, was reported by Vasey from Illinois in 1861 but, as no specimens have been discovered from Illinois, I assume Vasey's report was in error.

Potamogeton capillaceus Poir. Kibbe reported this essentially Coastal Plain species from Hancock County in 1952, but the specimen on which this identification was based is *P. diversifolius* Raf.

Potamogeton compressus L. Patterson (1876) misunderstood this species when he reported it from Illinois, confusing it with *P. friesii* Rupr.

Potamogeton dimorphus Raf. Pepoon's reference (1927) to this species from Illinois is actually meant for *P. diversifolius* Raf., since these two species were frequently confused early in the twentieth century.

Potamogeton fluitans Roth. Mead, who first recorded this binomial from Illinois in 1846, and other Illinois botanists were actually misidentifying specimens of *P. nodosus* Poir.

Potamogeton heterophyllus Schreb. Schreber's binomial has frequently been misapplied to Illinois specimens of *P. gramineus* L., but actually it refers to an entirely different species unknown from the eastern United States.

Potamogeton hybridus Petagna. Lapham (1857) first recorded this species from Illinois, and was followed by both Patterson (1874) and Brendel (1887). The material on which their identifications were made is really *P. diversifolius* Raf.

Potamogeton interruptus Kitaibel in Schultes. Specimens on which the report, in 1927 by Pepoon, of this species from Illinois was based were collected by E. J. Hill in 1881 in South Chicago. Since the specimens are sterile, it is not possible to identify them positively, but they probably represent *P. pectinatus* L., a species with which *P. interruptus* often is combined.

Potamogeton lonchites Tuckerm. The various reports of this species from Illinois by Patterson (1876), Schneck (1876), and Brendel (1887) would have been for *P. nodosus* Poir.

Potamogeton lucens L. There has been much confusion regarding this binomial which refers to a species generally restricted to the Old World. The Illinois specimens which have been called this are *P. illinoensis* Morong.

Potamogeton perfoliatus L. This species, restricted to boreal North America and the Old World, has been confused by Patterson (1876) for *P. richardsonii* (Benn.) Rydb.

Potamogeton spirillus Tuckerm. Reported by both Patterson (1876) and Brendel (1887), this is one of several species confused with *P. diversifolius* Raf. and erroneously attributed to Illinois. It should be pointed out, however, that the range of *P. spirillus* includes Illinois and it may well be that this species occurs somewhere in the state.

Potamogeton zosterifolius Schum. Until Fernald (1932) pointed out that *P. zosterifolius* was entirely different from our American material, this binomial was used for what is now *P. zosteriformis* Fern.

Sparganium angustifolium Michx. This species was originally attributed to Illinois in 1944 by Muenscher and later recorded by Fernald (1950) from Illinois. I have not been able to locate any Illinois specimens of this species.

Sparganium natans L. Vasey's report in 1861 of this species from Illinois is apparently in error.

Vallisneria spiralis L. In earlier days, distinction was not made between Linnaeus' *V. spiralis* and the specimens of *Vallisneria* found in eastern North America. It is now clear that the American plants are distinct and should be known as *V. americana* Michx.

Wolffia brasiliensis Wedd. This species does not occur in Illinois, although attributed to this state by Patterson (1876),

Brendel (1887), and Hegelmaier (1896). Since there is morphological similarity between this species and *W. punctata,* the references must have been intended for *W. punctata.*

Xyris flexuosa Muhl. ex Ell. This binomial was used by most botanists erroneously for *Xyris torta* Sm. in Rees. It actually should be restricted to a different species of the Coastal Plain.

Zigadenus elegans Pursh. Reference to this western white camas by Pepoon in 1909 is an error for *Z. glaucus* Nutt.

Summary of the Taxa Treated in This Volume

Orders and Families	Genera	Species	Varieties
Order 1. **Alismales**			
Family 1. Butomaceae	1	1	
Family 2. Alismaceae	3	12	
Family 3. Hydrocharitaceae	3	5	
Order 2. **Zosterales**			
Family 4. Juncaginaceae	2	3	
Family 5. Potamogetonaceae	1	23	
Family 6. Ruppiaceae	1	1	
Family 7. Zannichelliaceae	1	1	
Order 3. **Najadales**			
Family 8. Najadaceae	1	5	
Order 4. **Arales**			
Family 9. Araceae	5	6	1
Family 10. Lemnaceae	4	14	
Order 5. **Typhales**			
Family 11. Sparganiaceae	1	5	
Family 12. Typhaceae	1	2	
Order 6. **Commelinales**			
Family 13. Xyridaceae	1	2	
Family 14. Commelinaceae	2	8	
Family 15. Pontederiaceae	3	4	
Family 16. Juncaceae	2	24	4
Totals	32	116	5

APPENDIX: ADDITIONS AND CHANGES TO THE FIRST EDITION

Since the publication of the first edition of Flowering Rush to Rushes in 1970, several additions to the flora of Illinois in the families covered by this book have been made, and many new distributional records have been added. Also, a large number of nomenclatural revisions have taken place, resulting in numerous alterations in scientific names. All of the changes are reflected in the entries below.

Since the publication of the first edition of this work, a few new families of monocots have been added to the Illinois flora, and the names of others have changed. A new key to the monocots of Illinois is provided. An asterisk following the name of the family indicates that that family is treated in a separate volume of The Illustrated Flora of Illinois series. An asterisk following the name of a species indicates that there is no known specimen from Illinois.

KEY TO THE FAMILIES OF MONOCOTS IN ILLINOIS

1. Plants climbing or twining (if erect, then usually with a few weak tendrils from the upper axils); leaves net-veined; flowers unisexual.
    2. Inflorescence umbellate; ovary superior; fruit a berry_____Smilacaceae*
    2. Inflorescence glomerulate or paniculate; ovary inferior; fruit a capsule_____
    _____Dioscoreaceae*
1. Plants erect or floating in water (tendrils never present); leaves mostly parallel-veined; flowers bisexual or unisexual.
    3. Plants with 1 or 2 whorls of leaves.
        4. Flowers actinomorphic; ovary superior; stamens 6.
            5. Plants never more than 50 cm tall; flower usually borne singly_____
            _____Trillium and Medeola in Liliaceae*
            5. Plants more than 50 cm tall; flowers usually more than 1_____
            _____Lilium in Liliaceae*
        4. Flowers zygomorphic; ovary inferior; stamen 1____Isotria in Orchidaceae*
    3. Plants with leaves alternate, opposite, basal, or absent.
        6. Flowers crowded together in a spadix.
            7. Spathe at least partially surrounding the spadix_____Araceae
            7. Spathe absent, the spadix borne naked_____Acoraceae

6. Flowers not crowded into a spadix (in *Ruppia*, 2 flowers are borne in a spadixlike structure).
  8. Plants thalloid, floating in water_____Lemnaceae
  8. Plants with roots, stems and usually leaves, aquatic or terrestrial.
    9. Perianth absent or reduced to very minute scales (lodicules) or bristles.
      10. Each flower subtended by 1 or more scales; plants generally not true aquatics.
        11. Leaves 2-ranked; sheaths usually open; stems usually hollow with solid nodes, often terete; anthers attached above the base_
          _____Poaceae*
        11. Leaves (when present) 3-ranked; sheaths closed; stems solid with soft nodes, often 3-angled; anthers attached at the base__
          _____Cyperaceae*
      10. Flowers not subtended by individual scales; plants mostly aquatic.
        12. Plants erect; inflorescence terminal, spicate, thick; leaves very long, linear, strap-shaped_____ Typhaceae
        12. Plants not erect, free-floating or sometimes rooted in bottom mud; inflorescence axillary or terminal and slenderly spicate; leaves not very long, usually not linear, usually not strap-shaped.
          13. Leaves alternate; stamens 2 or 4; inflorescence spicate and usually terminal or with flowers borne in pairs in a spadixlike structure.
            14. Stamens 2; flowers on a short spadixlike structure, concealed within the leaf sheath; fruit stipitate, drupelike__
              _____Ruppiaceae
            14. Stamens 4; flowers in a spike or head; fruit sessile, appearing as an achene upon drying_____
              _____Potamogetonaceae
          13. Leaves opposite; stamen 1; inflorescence not spicate, axillary.
            15. Carpel 1; fruit beakless_____ Najadaceae
            15. Carpels 2–4; fruit beaked_____ Zannichelliaceae
    9. Perianth present, composed of either calyx or corolla or both (plants with perianth reduced to minute scales or bristles should be sought under the first 9).
      16. Pistils simple, more than 1, separate or slightly coherent at base.

17. Calyx and corolla differentiated in color and texture.
 18. Inflorescence umbellate; pistils 6, coherent at base; fruit a follicle_____Butomaceae
 18. Inflorescence not umbellate; pistils 10 or more, free to base; fruit an achene_____Alismataceae
17. Calyx and corolla not differentiated in color and texture.
 19. Carpels usually 6, the axis between them slender_____
  _____Juncaginaceae
 19. Carpels 3, the axis between them broadly 3-winged_____
  _____Scheuchzeriaceae
16. Pistil 1, compound.
 20. Ovary superior.
  21. Calyx and corolla differentiated in color and texture.
   22. Flowers crowded together in a dense head; leaves basal_
    _____Xyridaceae
   22. Flowers borne in cymes or umbels; leaves cauline_____
    _____Commelinaceae
  21. Calyx and corolla undifferentiated in color and texture.
   23. Flowers unisexual.
    24. Leaves net-veined; flowers in umbels_____
     _____Smilacaceae[*]
    24. Leaves parallel-veined; flowers in globose clusters, racemes, or panicles.
     25. Perianth small, greenish; flowers aggregated in dense globose clusters; stamens 3_____
      _____Sparganiaceae
     25. Perianth usually conspicuous, greenish, yellowish, white, or bronze-purple; flowers in racemes or panicles; stamens 6_____Liliaceae[*]
   23. Flowers bisexual.
    26. Perianth scarious_____Juncaceae
    26. Perianth petaloid.
     27. Stamens 3_____Pontederiaceae
     27. Stamens 6 (or 4).
      28. Stamens of different sizes_____
       _____Pontederiaceae
      28. Stamens all alike.
       29. Leaves evergreen, rigid; stems woody__
        _____*Yucca* in Liliaceae[*]

29. Leaves deciduous, mostly not rigid; stems herbaceous_____Liliaceae*

20. Ovary inferior.

30. Plants growing in water.

31. Leaves whorled_____Hydrocharitaceae

31. Leaves basal, or cauline and alternate.

32. Stamens 2 or 6–12, never 3; flowers unisexual; styles not petaloid_____Hydrocharitaceae

32. Stamens 3; flowers bisexual; styles petaloid_____ _____Iridaceae*

30. Plants growing on land.

33. Flowers zygomorphic; stamens 1 or 2 or composed of only a single anther half.

34. Leaves pinnately veined; stamen reduced to ½ an anther_____Marantaceae*

34. Leaves parallel-veined; stamens 1 or 2_____ _____Orchidaceae*

33. Flowers actinomorphic or nearly so; stamens 3 or 6.

35. Stamens 3; styles sometimes petaloid_____ _____Iridaceae*

35. Stamens 6; styles not petaloid.

36. Leaves reduced to scales; plants lacking chlorophyll, at most 3 cm tall_____Thismiaceae*

36. Leaves blade-bearing; plants with chlorophyll, well over 3 cm tall.

37. Flowers greenish yellow; at least some of the leaves more than 2 cm broad, fleshy_____ _____Manfreda in Agavaceae*

37. Flowers not greenish yellow; none of the leaves as much as 2 cm broad, not fleshy__ _____Amaryllidaceae*

Page 26. *Echinodorus tenellus* (Mart.) Buchenau var. *parvulus* (Engelm.) Fassett. Add the following county to the map on page 26: Mason.

Page 27. *Echinodorus berteroi* (Spreng.) Fassett var. *lanceolatus* (Wats. & Coult.) Fassett. The following counties should be added to the map on page 28: Adams, Bureau, Calhoun, Fulton, Knox, Lawrence, Macon, Mason, Peoria, Perry.

Page 29. *Echinodorus cordifolius* (L.) Griseb. The following counties should be added to map on page 29: DeWitt, Henderson, Morgan, Peoria, Piatt, Pulaski.

Page 31. Because of the addition of three species of *Sagittaria* to the Illinois flora since the publication of the first edition of *Flowering Rush to Rushes,* a new key is provided:

KEY TO THE SPECIES OF Sagittaria IN ILLINOIS

1. Pedicels of fruits recurved, rarely spreading; sepals of pistillate flowers mostly erect.
    2. Fruiting heads 1.2–2.0 cm in diameter; leaves hastate to sagittate_____ _____*S. calycina*
    2. Fruiting heads up to 1.2 cm in diameter; leaves linear-ovate to lance-elliptic_ _____*S. platyphylla*
1. Pedicels of fruits spreading to ascending or absent; sepals of pistillate flowers mostly spreading to recurved.
    3. Filaments pubescent.
        4. Pistillate flowers sessile or nearly so_____*S. rigida*
        4. Pistillate flowers pedicellate.
            5. Rhizomes present; stolons and corms absent_____*S. graminea*
            5. Rhizomes absent; stolons and corms present_____*S. cristata*
    3. Filaments glabrous.
        6. Emersed leaves linear to ovate, tapering or rounded at base, rarely sagittate.
            7. Leaves lanceolate to ovate; emersed plants with erect to ascending petioles_____*S. ambigua*
            7. Leaves linear to occasionally sagittate; emersed plants with recurved petioles_____*S. cuneata*
        6. Emersed leaves cordate, sagittate or hastate.
            8. Bracts free or connate for less than ¼ their length.
                9. Petioles winged in cross-section; beak of achene strongly recurved__ _____*S. australis*
                9. Petioles ridged in cross-section; beak of achene ascending, not recurved_____*S. brevirostra*
            8. Bracts connate for at least ¼ their length.
                10. Beak of achene 1–2 mm long, horizontal_____*S. latifolia*
                10. Beak of achene less than 1 mm long, erect or curved_____*S. cuneata*

Page 32. *Sagittaria calycina* Engelm. Add the following counties to the map on page 34: Carroll, Cass, Grundy, Hancock, Macoupin, Madison, Morgan, Saline, Schuyler, St. Clair, Union, Washington, Whiteside, Williamson.

Page 34. *Sagittaria rigida* Pursh. Add the following county to the map on page 36: Kane.

Page 36. *Sagittaria graminea* Michx. Add the following counties to the map on page 37: Bureau, DeKalb, Johnson, Kane, Mason, Will.

Page 39. *Sagittaria cuneata* Sheldon. Add the following counties to the map on page 41: Coles, DuPage, Hancock, Marshall, Menard, Will, Winnebago.

Page 41. *Sagittaria brevirostra* Mack. & Bush. Add the following counties to the map on page 43: Coles, DeKalb, Edwards, Union.

Page 43. *Sagittaria longirostra* (Micheli) J. G. Sm. The correct binomial for this species appears to be **Sagittaria australis** (J. G. Sm.) Small. Its taxonomy follows:

Sagittaria australis (J. G. Sm.) Small, Fl. SE. U.S. 46. 1903.
Sagittaria sagittifolia var. *longirostra* Micheli in DC. Monogr. Phan. 3:69. 1881, misapplied.
Sagittaria longirostra (Micheli) J. G. Sm. Mem. Torrey Club 5:26. 1894, misapplied.
Sagittaria longirostra (Micheli) J. G. Sm. var. *australis* J. G. Sm. in Mohr, Bull. Torrey Club 24:20. 1897.
Sagittaria engelmanniana J. G. Sm. ssp. *longirostra* (Micheli) Bogin, Mem. N.Y. Bot. Gard. 9:223. 1955.

The following counties should be added to the map on page 45: Pulaski, Union.

Page 45. *Sagittaria latifolia* Willd. The following counties should be added to the map on page 47: Clinton, Coles, Franklin, Gallatin, Hardin, Iroquois, Knox, Logan, Perry, Pulaski, Randolph, Washington.

After *Sagittaria latifolia* Willd., add the following species of *Sagittaria* to the Illinois flora:

Sagittaria ambigua J. G. Sm. N. Am. Sagittaria 22–23, t. 17. 1894. Fig. A1.

Plants monoecious or dioecious, erect, to 50 cm tall, the flowering stem equaling or slightly shorter than the leaves; petiole slender, rounded, with expanded base; blade ovate to lanceolate, never sagittate, prominently nerved, acute to acuminate, tapering or slightly rounded at the base, to 22 cm long, to 7 cm broad; peduncle slender, erect, unbranched, the flowers in whorls of 3, the upper whorls staminate or pistillate, the lower 1–5 whorls pistillate; bracts 3, opposite the pedicel, usually free to the

A1. Sagittaria ambigua (Plains Arrowhead). a. Habit. b. Achene.

base, narrowly lanceolate, to 2 cm long; pedicel slender, terete, to 4 cm long; sepals 3, basally connate, short, green, separate, persistent, reflexed in fruit, minutely papillose, ovate-lanceolate, acute, 5–7 mm long, 1–3 mm broad; petals 3, free, white, a little longer than the sepals, caducous, narrowing to a slender claw at base, to 2 cm long, sometimes nearly as broad; stamens numerous, the filaments to 3 mm long, not dilated at base, glabrous, shorter than the anther; fruiting head globoid, up to 8 mm in diameter when mature; achenes obovoid, 1.5–2.0 mm long, 1 mm broad, with thin narrow dorsal and ventral wings, the beak to 0.2 mm long, triangular and spreading at a right angle to the body of the fruit.

COMMON NAME: Plains Arrowhead.

HABITAT: Ditches, borders of lakes and ponds, sloughs, often in standing water.

RANGE: Illinois, Indiana, Kansas, Missouri, Oklahoma.

ILLINOIS DISTRIBUTION: Adams, Carroll, and Pike counties.

This species is found in wetlands usually in areas of prairies. It is one of the species of *Sagittaria* that does not form sagittate leaves. It differs from specimens of *S. cuneata,* which lack sagittate leaves, by its erect, rather than recurved, peduncles. This species flowers from June to August.

Sagittaria cristata Engelm. Proc. Davenp. Acad. Nat. Sci. 4:29. 1883. Fig. A2.

Sagittaria graminea Michx. var. *cristata* (Engelm.) Bogin, Mem. N.Y. Bot. Gard. 9 (2):210. 1955.

Perennials with rhizomes but without stolons and corms; plants monoecious or dioecious, erect, 5–50 cm tall, the flower stalk equaling or slightly shorter than the leaves, rather slender in appearance; petiole long, slender, minutely ridged, with expanded base; blade either entire, hastate, or represented by bladeless phyllodia, prominently nerved, the unlobed blade linear-lanceolate, tapering to the acute apex, 2–17 cm long, 3–5 cm broad, the hastate blade with lobes usually unequal in both length and width, up to 2.5 cm long, about 2 cm broad, the phyllodia linear-lanceolate, tapering to the acuminate tip, to 19 cm long, to 2 cm broad; peduncle slender, minutely ridged, sparsely flowered, the flowers in whorls of 3, the upper flowers either staminate or pistillate, the lower flowers pistillate; bracts 3, opposite the pedicel, basally connate to fused one-half their entire length, weak, prominently nerved, hyaline in appearance, inconspicuous, ovate, obtuse, 2 mm long, 1–2 mm broad; pedicels long, slender, terete, with staminate and pistillate being the same length, 1–3 cm long; sepals basally connate, short, with a hyaline margin, persistent, reflexed in fruit, ovate-lanceolate with obtuse apex, 3–6 mm long, 1–3 mm broad; petals free, greatly exceeding the sepals, ovate, narrowing to a slender claw at the base, obtuse, 1–2 cm long, 1–2 cm broad; stamens numerous, the filaments up to 3 mm long, inflated,

A2. Sagittaria cristata (Crested Arrowhead). a. Staminate plant. b. Pistillate plant. c. Achene.

scaly, shorter than the anther; fruiting heads globoid, small, 4–8 mm in diameter when mature, smooth in appearance; achenes narrowly ovoid, 2 mm long, 1 mm broad, with both ventral and dorsal wings well developed, the dorsal one strongly scalloped, with 1–3 facial wings with one well developed, the beak usually oblique to horizontal and borne below the summit of the achene, 0.4–0.7 mm long.

COMMON NAME: Crested Arrowhead.

HABITAT: Borders of lakes and ponds.

RANGE: Ontario and Michigan to Minnesota, south to Nebraska and Missouri.

ILLINOIS DISTRIBUTION: Known from Carroll, JoDaviess, and Whiteside counties.

This species of the northern Midwest is distinguished from the similar *S. graminea* by the longer beak of the achene and the scalloped dorsal ridge on the achene. Submersed leaves are very stiff in *S. cristata,* while they are flaccid in *S. graminea.* It flowers from June to September.

Sagittaria platyphylla (Engelm.) J. G. Sm. N. Am. Sagittaria 29. 1894. Fig. A3.
Sagittaria graminea Michx. var. *platyphylla* Engelm. Man. Bot. N. U.S., ed. 5, 494. 1867.

Rhizomatous perennial, monoecious or dioecious; flowering stems to 75 cm tall, glabrous; leaves 10–70 cm long, glabrous, the petioles terete to flattened, the submerged leaves linear or absent, the emergent leaves lanceolate to ovate, occasionally with two basal lobes much shorter than the terminal lobe; flowering stems unbranched, erect; bracts connate about one-half their length, the apex obtuse, 3–7 mm long, about as broad; lower 1–3 whorls of flowers reflexed, pistillate, the pedicels thickened, glabrous, 15–30 mm long, the upper whorls of flowers staminate; sepals short, free, reflexed in fruit, 3–5 mm long; petals free, 5–10 mm long; stamens numerous, the filaments slender but somewhat swollen at the base, papillose, shorter than the anthers; fruiting heads globose, 0.9–1.2 cm in diameter when mature; achenes obovate, 1.5–2.0 mm long, glabrous, the beak curved, slender, 0.3–0.6 mm long.

COMMON NAME: Flat-leaved arrowhead.

HABITAT: Sloughs, ditches, around ponds, sometimes in standing water.

RANGE: West Virginia to Illinois and Missouri, south to Texas and northern Florida.

ILLINOIS DISTRIBUTION: Known only from St. Clair County.

Only *S. platyphylla* and *S. calycina* in Illinois have the pedicels of the fruits recurved and the sepals of the pistillate flowers mostly erect. *Sagittaria platyphylla* differs from *S. calycina* by its smaller fruiting heads up to 1.2 cm across and its non-sagittate leaves. It flowers from June to September.

Page 48. *Alisma plantago-aquatica* L. var. *americana* Roem. & Schultes. This taxon should be called **Alisma triviale** Pursh. Its taxonomy follows:

A3. Sagittaria platyphylla (Flat-leaved Arrowhead). Habit.

Alisma triviale Pursh, Fl. Am. Sept. 1:252. 1814.
Alisma plantago-aquatica L. var. *americanum* Roem. & Schultes, Syst. 7:1598. 1830.
Alisma plantago Bigel var. *triviale* (Pursh) BSP. Prel. Cat. N.Y. Pl. 58. 1888.
Alisma plantago-aquatica L. var. *triviale* (Pursh) Farwell, Ann. Rep. Comm. Parks & Boulev. Det. 11:44. 1900.

The following counties should be added to the map on page 50: Carroll, DeKalb, Fulton, Henderson, Henry, Iroquois, LaSalle, Rock Island.

Page 50. *Alisma subcordatum* Raf. The following counties should be added to the map on page 52: Bond, Effingham, JoDaviess, Knox, Lee, Ogle, Shelby.

Page 53. *Elodea densa* (Planch.) Caspary. This species is now known as **Egeria densa** Planch. Its taxonomy follows:

Egeria densa Planch. Ann. Sci. Nat. Bot. III, 11:80. 1849.
Elodea densa (Planch.) Caspary, Monaster. Kgl. Preuss. Akad. Wissensch. 1857:48. 1857.
Anacharis densa (Planch.) Vict. Contr. Lab. Bot. Univ. Montreal 18:41. 1931.

The following counties should be added to the map on page 53: Edwards, Jefferson.

Page 55. *Elodea canadensis* Rich. The following counties should be added to the map at the top of page 56: Cass, Crawford, DeKalb, Jackson, JoDaviess, Lawrence, Pulaski, Williamson.

Page 56. *Elodea nuttallii* (Planch.) St. John. The following counties should be added to the map at the bottom of page 56: Boone, Bureau, Carroll, Cass, Clark, Coles, Greene, Grundy, Hamilton, Iroquois, Jersey, Knox, Lake, Lawrence, Lee, Madison, McDonough, Moultrie, Pulaski, Rock Island, Schuyler, Shelby, Stephenson, Union, White.

Page 58. *Vallisneria americana* Michx. The following counties should be added to the map at the top of page 59: Alexander, Carroll, JoDaviess, Rock Island.

Page 59. *Limnobium spongia* (Boxc) Steud. The following counties should be added to the map at the bottom of page 59: Jackson, Pulaski.

Page 61. Since the genus *Scheuchzeria* is now placed in its own family, the Scheuchzeriaceae, a new key to the families of the Order Zosterales is provided here:

KEY TO THE FAMILIES OF THE ORDER Zosterales IN ILLINOIS

1. Perianth composed of two series of three members each; stamens 6; carpels 3 or 6; rooted perennials.
    2. Inflorescence a spikelike raceme, without bracteoles; leaves all basal_____
    _____Family Juncaginaceae
    2. Inflorescence loosely racemose, with bracteoles; leaves basal as well as cauline and alternate_____Family Scheuchzeriaceae
1. Perianth absent (4 sepal-like structures present in Potamogetonaceae); stamens 1, 2, or 4; carpels 2–4; free-floating aquatics.
    3. Leaves alternate (the uppermost rarely opposite); stamens 2 or 4; fruit not beaked, or obscurely beaked.
        4. Flowers in a spike or head; stamens 4; fruit appearing as an achene upon drying, sessile_____Family Potamogetonaceae
        4. Flowers on a short spadix concealed within the leaf sheath; stamens 2; fruit a drupe, stipitate_____Family Ruppiaceae
    3. Leaves opposite; stamen 1; fruit strongly beaked_____
    _____Family Zannichelliaceae

Page 62. *Triglochin maritimum* L. should be **Triglochin maritima** L. Add the following counties to the map on page 62: Tazewell, Will.

Page 62. *Triglochin palustre* L. should be **Triglochin palustris** L. Add the following counties to the map at the top of page 65: Kendall, Will.

Page 65. The genus *Scheuchzeria* is now placed in the family Scheuchzeriaceae.

SCHEUCHZERIACEAE—ARROW-GRASS FAMILY

Only the following genus is in this family.

Scheuchzeria L.—Arrow-grass

Plants with creeping rhizomes; leaves terete, linear, most of them basal; flowers perfect; sepals 3, greenish, deciduous; petals 3, greenish, persistent; stamens 6; carpels free except at base, developing into 3 follicles; ovules 2; inflorescence loosely racemose, bracteate.

This genus is sometimes placed in the Juncaginaceae, but it differs from that family by its loosely racemose inflorescence that is subtended by bracts and by its basal and cauline leaves.

Page 65. *Scheuchzeria palustris* L. var. *americana* Fern. There are no new records for this taxon in Illinois.

Page 67. Since *Potamogeton pectinatus* L. is now placed in the genus **Stuckenia,** a key to the two genera of Potamogetonaceae in Illinois follows:

1. Leaves not septate, some or all over 0.5 mm broad; peduncles rigid_____
_____1. *Potamogeton*
1. Leaves septate, 0.3–0.5 (-2.0) mm broad; peduncles flexible_____
_____2. *Stuckenia*

Page 70. *Potamogeton pectinatus* L. is now called **Stuckenia pectinata** (L.) Borner and is the only Illinois member of the genus *Stuckenia.*

Stuckenia Borner—Linear Pondweeds

Rhizomes present; stems terete; leaves submersed, sessile, linear, tapering to the base, obtuse to acute at the apex, the margins entire, 1- to 5-nerved; stipules adnate to the base of the leaves; spikes capitate or cylindric, submersed, on flexible peduncles; fruits rounded, with or without a beak, turgid.

The species is this genus usually have been included in the genus *Potamogeton. Stuckenia* species differ by their submersed fruits on flexible peduncles.

Six species comprise the genus, found in most parts of the world. Only the following is in Illinois.

Stuckenia pectinata (L.) Borner, Fl. Deut. Volk 713. 1912.
Potamogeton pectinatus L. Sp. Pl. 1:127. 1753.

The following counties should be added to the map on Page 70: Bureau, Cass, Champaign, Clark, Crawford, DeKalb, Douglas, DuPage, Ford, Greene, Henderson, Iroquois, Kane, Kendall, Knox, Lee, Logan, Macoupin, Madison, McLean, Mercer, Montgomery, Moultrie, Ogle, Perry, Piatt, Pike, Randolph, Richland, Rock Island, Shelby, Stark, Stephenson, Union, Warren, Whiteside, Will, Williamson, Woodford.

Since one additional species of *Potamogeton* has been added to the Illinois flora, and since *Potamogeton pectinatus* has been removed to the genus *Stuckenia,* a new key to the Illinois species of *Potamogeton* follows:

KEY TO THE SPECIES OF **Potamogeton** IN ILLINOIS

1. Leaves uniform.

2. Stipules adnate to leaf base; leaves auriculate at base_____2. *P. robbinsii*

2. Stipules free from leaf base; leaves rounded, cordate, or tapering at base, never auriculate.

   3. Leaves sharply serrulate, crispate; fruits 5–6 mm long____3. *P. crispus*

   3. Leaves entire or minutely denticulate, usually not crispate; fruits 1.6–5.0 mm long.

      4. Leaves 5–30 mm broad, rounded or cordate at base.

         5. Leaves entire; fruits 4–5 mm long, keeled_____4. *P. praelongus*

         5. Leaves minutely denticulate; fruits 2.5–3.5 mm long, not keeled___

         _____5. *P. richardsonii*

      4. Leaves up to 5 mm broad, tapering to base.

         6. Leaves with 15–35 nerves; spikes with 6–11 whorls of flowers; stems flattened; fruits quadrate, 4–5 mm long_____6. *P. zosteriformis*

         6. Leaves with 3–7 nerves; spikes with 1–5 whorls of flowers; stems not flattened; fruit obovoid to ovoid, 1.9–3.0 mm long.

           7. Stipules connate, forming cylinders.

              8. Spikes subcapitate, 2–5 mm long; fruits keeled; sepaloid connectives 0.5–1.0 mm long; leaves glandless at base_____

              _____7. *P. foliosus*

              8. Spikes elongate, more or less interrupted, 6–15 mm long; fruits rounded on back; sepaloid connectives 1.2–2.5 mm long; leaves biglandular at base.

                 9. Leaves 5- to 7-nerved, thin, translucent_____8. *P. friesii*

                 9. Leaves 3-nerved, firm, opaque.

                    10. Stipules chartaceous, white, becoming fibrous_____

                    _____9. *P. strictifolius*

                    10. Stipules membranous, olive, not becoming fibrous_____

                    _____10. *P. pusillus*

           7. Stipules free_____11. *P. berchtoldii*

1. Leaves of two kinds, the floating usually shorter and broader than the submersed, or floating leaves sometimes absent.

11. Leaves entire.

   12. Submersed leaves up to 2 mm broad, 1- to 5-nerved.

      13. Spikes uniform; stipules free from base of leaf; fruits 1.6–5.0 mm long.

         14. Submersed leaves 0.1–0.5 mm broad, 1- to 3-nerved; fruits 1.6–2.2 mm long_____12. *P. vaseyi*

         14. Submersed leaves 0.5–2.0 mm broad, 3- to 5-nerved; fruits 3–5 mm long_____13. *P. natans*

13. Spikes of two or more kinds, those in the axils of the submersed leaves subgloboid, those in the axils of the floating leaves elongate; stipules adnate to base of leaf; fruits 1.0–1.5 mm long.

15. Fruits beaked, with sharp-tipped keels; floating leaves obtuse to acute_____14. *P. diversifolius*

15. Fruits beakless, without sharp-tipped keels; floating leaves acute to acuminate_____*P. bicupulatus* (see below)

12. Submersed leaves 5–75 mm broad, 7- to 21-nerved.

16. Stems compressed; submersed leaves linear, 5–10 mm broad; fruit capped with a minute, toothlike beak_____15. *P. epyhydrus*

16. Stems terete; submersed leaves lanceolate to ovate, 10–75 mm broad; fruit capped with a prominent beak.

17. Fruiting spike 4 cm long or longer; stipules of submersed leaves subpersistent; fruit tapering to base; some nerves of floating leaves more prominent than other nerves_____
_____16. *P. amplifolius*

17. Fruiting spike 2.0–3.5 cm long; stipules of submersed leaves falling away early; fruit rounded at base; all nerves of floating leaves uniform_____17. *P. pulcher*

11. Leaves minutely denticulate.

18. Fruits (including beak) 3.5–4.3 mm long_____18. *P. nodosus*

18. Fruits (including beak) 1.7–3.5 mm long.

19. Stems simple or once-branched; stipules strongly keeled.

20. Floating leaves 2.0–6.5 cm wide; fruits normally developed___
_____19. *P. illinoensis*

20. Floating leaves 0.5–1.0 cm wide; fruits not maturing_____
_____20. *P. X hagstromii*

19. Stems repeatedly branched; stipules faintly keeled.

21. Floating leaves elliptic to ovate, the petioles usually as long as or longer than the blades; stipules obtuse.

22. Submersed leaves acute at apex; fruit strongly 3-keeled, well-developed_____21. *P. gramineus*

22. Submersed leaves cuspidate at apex; fruit obscurely 3-keeled, poorly developed_____22. *P. X spathulaeformis*

21. Floating leaves oblong- to linear-lanceolate, the petioles shorter than the blades; stipules acuminate_____23. *P. X rectifolius*

Page 72. *Potamegeton robbinsii* Oakes. Add the following counties to the map on page 72: Cook, McHenry.

Page 74. *Potamogeton crispus* L. Add the following counties to the map on page 74: Adams, Boone, Coles, Cumberland, Edwards, Ford, Fulton, Greene, Hancock, Henry, Jackson, Jersey, JoDaviess, Johnson, Kane, Knox, Lawrence, Lee, Madison, McHenry, Mercer, Pulaski, Randolph, Rock Island, Sangamon, Shelby, Washington, Will, Williamson, Woodford.

Page 74. *Potamogeton praelongus* Wulfen. There are no new records for this species.

Page 77. *Potamogeton richardsonii* (Benn.) Rydberg. Add the following county to the map at the bottom of page 77: Kankakee.

Page 79. *Potamogeton zosteriformis* Fern. Add the following county to the map on page 79: Peoria.

Page 79. *Potamogeton foliosus* Raf. Add the following counties to the map on page 82: Alexander, Brown, Bureau, Carroll, Christian, Clark, Clinton, Crawford, DeKalb, DeWitt, Edwards, Effingham, Fayette, Franklin, Greene, Grundy, Hancock, Iroquois, Jefferson, Kane, Kendall, LaSalle, Lawrence, Livingston, Marion, Mercer, Monroe, Montgomery, Morgan, Perry, Pike, Pope, Randolph, Rock Island, Saline, Schuyler, Shelby, Warren, Washington, Will, Williamson.

Page 82. *Potamogeton friesii* Rupr. Add the following counties to the map on page 84: McHenry, Shelby.

Page 84. *Potamogeton strictifolius* Benn. Add the following county to the map at the top of page 86: Lake.

Page 86. *Potamogeton pusillus* L. The following counties should be added to the map at the bottom of page 86: Adams, Alexander, Calhoun, Cass, Coles, Cook, Crawford, Cumberland, DeWitt, Douglas, DuPage, Edgar, Edwards, Effingham, Fayette, Ford, Franklin, Grundy, Hamilton, Henderson, Jackson, Jasper, Jersey, JoDaviess, Johnson, Kane, Knox, LaSalle, Lawrence, Lee, Macon, Madison, Marion, McLean, Mercer, Monroe, Morgan, Moultrie, Perry, Piatt, Pike, Randolph, Richland, Saline, Sangamon, Shelby, Vermilion, Warren, Washington, Wayne, Williamson.

Page 88. *Potamogeton berchtoldii* Fieber. There are no new records for this species.

Page 88. *Potamogeton vaseyi* Robbins. Add the following county to the map on page 90: Grundy.

Page 91. *Potamogeton natans* L. Add the following county to the map on page 91: Cook.

Page 91 *Potamogeton diversifolius* Raf. Add the following counties to the map on page 94: Alexander, Bond, Clark, Clay, Coles, Cook, Crawford, Cumberland, Edwards, Effingham, Franklin, Gallatin, Grundy, Hamilton, Hardin, Jasper, Lawrence, Mason, McDonough, Monroe, Montgomery, Perry, Pike, Randolph, Shelby, Wayne, White, Will.

Page 94. *Potamogeton epihydrus* Raf. There are no new records for this species.

Page 96. *Potamogeton amplifolius* Tuckerm. There are no new records for this species.

Page 98. *Potamogeton pulcher* Tuckerm. Add the following county to the map on page 99: Kane.

Page 100. *Potamogeton nodosus* Poir. Add the following counties to the map on page 100: Adams, Clark, Crawford, Cumberland, DeWitt, Edgar, Edwards, Ford, Hamilton, Jasper, Jefferson, Jersey, Johnson, Kane, Kendall, LaSalle, Lee, Marion, McLean, Mercer, Morgan, Moultrie, Pope, Randolph, Rock Island, Saline, Sangamon, Scott, Shelby, St. Clair, Washington, Whiteside.

Page 102. *Potamogeton illinoensis* Morong. Add the following counties to the map on page 102: JoDaviess, Kendall, Knox, Lawrence, Macoupin, McLean, Perry, St. Clair.

Page 104. *Potamogeton X hagstromii* Benn. There are no new records for this hybrid.

Page 105. *Potamogeton gramineus* L. Add the following counties to the map on page 106: Lake, McHenry.

Page 106. *Potamogeton X spathulaeformis* (Robbins) Morong. There are no new records for this hybrid.

Page 109. *Potamogeton X rectifolius* Benn. There are no new records for this hybrid.

After *Potamogeton X rectifolius* Benn, add the following species of *Potamogeton* that is new to Illinois:

Potamogeton bicupulatus Fern. Mem. Amer. Acad. Arts, n. s. 17:112. 1932. Fig. A4.
Potamogeton diversifolius Raf. var. *trichophyllus* Morong, Mem. Torrey Club 3 (2):49. 1893.

Rootstocks slender, fibrous; stems profusely branched, compressed, occasionally simple, slender; leaves dimorphic, although the floating ones sometimes absent; floating leaves elliptic, oval, or narrowly obovate, acute to long-tapering at the apex, rounded at the base, 1.2–3.5 cm long,

3–11 mm broad, 3- to 7-nerved, the compressed petiole 1.7–2.0 cm long, adnate to the stipules; stipules to 3 cm long, not becoming fibrous; submersed leaves linear, obtuse at the apex, to 45 mm long, 0.1–0.6 mm wide, sessile, lax, the stipules 2–15 mm long, sheathing for one-half of their length; spikes continuous, subgloboid to elongate, 0.6–1.2 (–2.0) cm long, those in the axils of the submersed leaves subgloboid, those in the axils of the floating leaves elongate; flowers sessile; sepaloid connectives suborbicular to reniform; fruits smooth and with one strong and two weak keels without sharp tips on the back, 1.1–2.1 mm long, 1.1–2.0 mm broad, beakless, green to greenish brown.

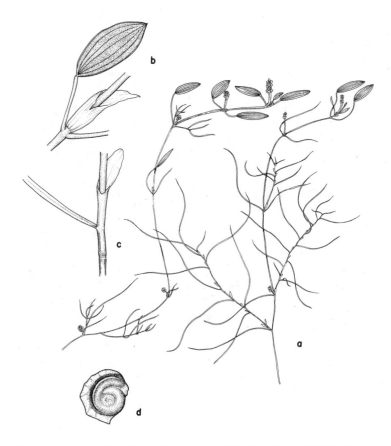

A4. Potamogeton bicupulatus (Slender-leaved Pondweed). a. Habit. b. Floating leaf with sheath. c. Submersed leaf with sheath. d. Achene.

COMMON NAME: Slender-leaved Pondweed.

HABITAT: Acid waters.

RANGE: Ontario to Wisconsin, south to Illinois, Pennsylvania, and Rhode Island.

ILLINOIS DISTRIBUTION: The only record for this species is from Wabash County.

This species in the past has usually been considered a variety of *P. diversifolius*. It differs from *P. diversifolius* by its acute to acuminate floating leaves and its beakless fruits without sharp-tipped keels.

Page 111. *Ruppia maritima* L. var. *rostrata* Agardh should now be called **Ruppia cirrhosa** (Petagna) Grande. Its taxonomy follows:

Ruppia cirrhosa (Petagna) Grande, Bull. Orto Bot. Regia Univ. Napoli 5:58. 1918.

Buccaferrea cirrhosa Petagna, Inst. Bot. 5:1826. 1787.

Ruppia maritima L. var. *rostrata* Agardh, Phsiog. Sallsk. Arbetr. 27. 1823.

Ruppia rostellata Koch ex Reichenb. Ic. Pl. Crit. 2:66. 1824.

There are no new records for this species.

Page 113. *Zannichellia palustris* L. The following counties should be added to the map on page 113: Adams, Cook, DeKalb, Hancock, Knox, Lake, Lee, McHenry, Morgan, Ogle, Rock Island, Sangamon, Schuyler, St. Clair, Stephenson, Vermilion, Warren, Whiteside, Woodford.

Page 116. *Najas flexilis* (Willd.) Rostk. & Schmidt. The following counties should be added to the map on page 116: Champaign, DuPage, Ford, Grundy, JoDaviess, Livingston, Shelby, Whiteside.

Page 116. *Najas guadalupensis* (Spreng.) Magnus. Two rather distinct subspecies of *N. guadalupensis* occur in Illinois. They may be distinguished by the following key:

1. Leaves with 50–100 teeth per side; stems 0.2–1.0 mm thick_____
_____*N. guadalupensis* ssp. *guadalupensis*
1. Leaves with 20–40 teeth per side; stems 1–2 mm thick_____
_____*N. guadalupensis* ssp. *olivacea*

Najas guadalupensis (Spreng.) Magnus, Beitr. Gatt. 8. 1870. ssp. **guadalupensis**
Caulinia guadalupensis Spreng. Syst. 1:20. 1825.

Naias flexilis (Willd.) Rostk. & Schmidt var. *guadalupensis* (Spreng.) A. Br. Journ. Bot. Brit. & For. 2:276. 1864.

Leaves with 50–100 teeth per side; stems 0.2–1.0 mm thick.

Najas guadalupensis (Spreng.) Magnus ssp. **olivacea** (C. Rosend. & Butters) Haynes & Hellq. Novon 6:371. 1996. Not illustrated.
Najas olivacea C. Rosend. & Butters, Rhodora 37:347–48. 1935.
Najas guadalupensis (Spreng.) Magnus var. *olivacea* (C. Rosend. & Butters) Haynes, Sida 8:43. 1979.

Leaves with 20–40 teeth per side; stems 1–2 mm thick.

COMMON NAME: Stout Naiad.
HABITAT: Ponds and shallow lakes.
RANGE: Quebec to Ontario, south to Iowa, Illinois, and New York.
ILLINOIS DISTRIBUTION: Scattered throughout the state; Adams, Alexander, Coles, Mason, Shelby, and Union counties.

This subspecies is relatively stout when compared to the typical subspecies.

Page 118. *Najas gracillima* (A. Br.) Magnus. Add the following counties to the map at the top of page 120: Crawford, Edwards, Franklin, Hardin, Jackson, Richland.

Page 120. *Najas minor* All. Add the following counties to the map at the bottom of page 120: Champaign, Coles, Cumberland, Douglas, Effingham, Hardin, Perry, Union.

Page 120. *Najas marina* L. There are no new records for this species.

Page 124. Araceae. The genus *Acorus* is now in its own family, the Acoraceae, while the genus *Calla* is a new genus for Illinois in the Araceae. A new key to the families Acoraceae and Araceae in Illinois is presented below:

1. True spathes absent; ovary 2- to 3-celled; fruit dry; leaves linear_____
_____*Acorus* in Acoraceae
1. Spathes present; ovary 1-celled; fruit fleshy; leaves broad_____
_____Family Araceae
  2. Some or all the flowers perfect; spadix globose or short-cylindric; plants with rhizomes; leaves simple, without basal auricles.
    3. Spadix globose; spathe more or less enclosing the spadix, not petaloid; perianth 4-parted; leaves at maturity at least 15 cm broad; plants foul-smelling_____*Symplocarpus*

3. Spadix short-cylindric; spathe broad, flat, petaloid, surrounding but not enclosing the spadix; perianth absent; leaves less than 15 cm broad; plants not foul-smelling_____*Calla*

2. All flowers unisexual; spadix elongated; plants with corms or fleshy roots; leaves compound or simple with basal auricles.

    4. Staminodia present; berries brownish or greenish; leaves simple, arrowhead-shaped.

        5. Plants rhizomatous; spathe tubular only at base; flowers confined to the lower half of the spadix_____*Arum*

        5. Plants with fleshy roots; spathe tubular at both ends, opening at the middle; flowers covering all or most of the spadix_____*Peltandra*

    4. Staminodia absent; berries red; leaves compound_____*Arisaema*

ACORACEAE—SWEET FLAG FAMILY

Characters of the genus.

This only genus in the family is often placed in the Araceae, but differs from the Araceae in lacking a spathe associated with the inflorescence and by its linear, grasslike leaves.

KEY TO THE SPECIES OF Acorus IN ILLINOIS

1. Midvein and 2–several lateral nerves prominent, not excentric_____
_____1. *A. americanus*

1. Only the midvein prominent and excentric, the lateral nerves obscure_____
_____2. *A. calamus*

    1. **Acorus americanus** (Raf.) Raf. New Fl. & Bot. N. Am. 1:57. 1836. Fig. A5.

    Acorus calamus L. var. *americanus* Raf. Med. Fl. 1:25. 1828.

Rhizome stout, creeping, aromatic; leaves numerous, basal, linear, to 1 (–2) m long, to 2.0 (–2.5) cm broad, with 2–several prominent parallel veins; scape flattened, leaflike, to 1 m long, extended up to 60 cm beyond the spadix as a modified spathe; inflorescence crowded into a dense spadix to 10 cm long, 0.8–2.0 cm thick; flowers small, brownish yellow, the perianth segments concave; fruits obpyramidal, indehiscent, fertile.

COMMON NAME: American Sweet Flag.

HABITAT: Marshes, ditches, fens, spring branches, sloughs, often in standing water.

RANGE: Labrador to Alaska, south to Washington, Nebraska, Illinois, and Virginia.

ILLINOIS DISTRIBUTION: Apparently confined to the northern one-fourth of the state; Carroll, DeKalb, Kane, Lake, and Will counties.

A5. Acorus americanus (American Sweet Flag). a. Habit. b. Spadix. c. Leaf with veins.

Although Rafinesque described this species as distinct from *A. calamus*, most botanists except Wilson (1960) and Thompson (1995) considered the two to be the same. *Acorus calamus*, however, is native to Asia and Europe but widely distributed in North America. *Acorus americanus* is readily distinguished by its two prominent, non-excentric veins in each leaf and by its fertile seeds. This species blooms from May to mid-August.

The leaves of *Acorus americanus* resemble those of some grasses, some sedges, irises, bur-reeds, and young cat-tails, but the strongly sweet-scented rhizomes are distinctive.

The rhizomes have a history of uses as a medicine and an additive to perfumes.

Page 125. *Acorus calamus* L. Add the following counties to the map on page 125: Clark, Cumberland, Douglas, Iroquois, Macoupin, Massac, Rock Island, Scott, St. Clair.

Page 127. *Symplocarpus foetidus* (L.) Nutt. Add the following counties to the map on page 128: Boone, Stephenson.

Page 128. *Arum italicum* Muhl. There are no new records for this species.

Page 130. *Peltandra virginica* (L.) Kunth. Add the following counties to the map on page 132: Cook, Jackson, Kane, Lee.

Page 132. *Arisaema dracontium* (L.) Schott. Add the following counties to the map on page 134: Bureau, Grundy.

Page 135. *Arisaema triphyllum* (L.) Schott var. *triphyllum*. There are no new records for this plant.

Page 135. *Arisaema triphyllum* (L.) Schott var. *pusillum* Peck. Add the following county to the map at the bottom of page 135: Jackson.

Page 137. After the illustrations of *Arisaema triphyllum*, add the following new genus and species for Illinois:

Calla L.—Water Arum

Characters of the species. This is the only species in the genus.

Calla palustris L. Sp. Pl. 2:968. 1753. Fig. A6.

Perennial with elongated rhizomes; leaves basal, ovate to nearly orbicular, short-acuminate at the apex, cordate at the base, entire, glabrous, 5–10 cm long, often nearly as broad, with curved-ascending, parallel veins; petioles rather stout, up to 20 cm long; spathe open, white, not concealing the spadix, ovate to elliptic, up to 6 cm long, narrowed to a linear, involute tip up to 10 mm long; spadix short-cylindric, 1.5–2.5 cm long, with a thickened stipe; flowers perfect; perianth absent; stamens 6, with flat, narrow filaments; berries red, 8–12 mm in diameter, with few seeds covered by a gelatinous material.

COMMON NAME: Wild Calla.
HABITAT: Swamps, bogs.
RANGE: Labrador to Alaska, south to British Columbia, Minnesota, northern Illinois, Pennsylvania, and New Jersey; also in Europe and Asia.
ILLINOIS DISTRIBUTION: Known only from Lake County.

A6. Calla palustris (Wild Calla). Habit.

The white spathes make this one of the most attractive native wetland plants in North America. The leaves are much more cordate than the leaves of *Alisma* species. *Caltha palustris,* the marsh marigold, has cordate leaves with teeth and net venation and yellow flowers. *Calla palustris* flowers from June to August.

Page 139. *Spirodela polyrhiza* (L.) Schleiden. Add the following counties to the map on page 139: Clark, Coles, Crawford, Cumberland, Douglas, Effingham, Gallatin, Jasper, JoDaviess, Lawrence, Marion, McDonough, Mercer, Pope, Pulaski, Putnam, Richland, Saline, Scott, Wabash.

Page 141. *Spirodela oligorhiza* (Kurtz) Hegelm. This species is now known as **Spirodela punctata** (Mey.) C. H. Thompson. Its taxonomy follows:

Spirodela punctata (G. Mey.) C. H. Thompson, Ann. Rep. Mo. Bot. Gard. 9:28. 1898.

Lemna punctata G. Mey. Prim. Fl. Ess. 262. 1818.

Lemna oligorhiza Kurtz, Journ. Linn. Soc. London 9:267. 1867.

Spirodela oligorhiza (Kurtz) Hegelm. Die Lemnaceen 147. 1868.

Landoldtia punctata (G. Mey.) D. H. Les & D. J. Crawford, Novon 9:530. 1999.

The following counties should be added to the map on page 141: Jackson, Lawrence.

Page 143. *Lemna* L. Since two additional species of *Lemna* have been added to the Illinois flora, a new key is provided below:

KEY TO THE SPECIES OF Lemna IN ILLINOIS

1. Fronds spatulate with long, persistent stipes, often submerged in compact masses; rootlet frequently absent_____7. *L. trisulca*
1. Fronds orbicular to elliptic, floating; rootlet present.
  2. Fronds producing rounded, rootless turions or specialized overwintering structures late in the year; fronds with a distinct line of papillae near apex_
    _____8. *L. turionifera*
  2. Fronds not producing rounded, rootless turions late in the year; fronds without a papilla or with indistinct lines of papillae.
    3. Fronds 3- or 5-nerved.
      4. Rootlet sheaths cylindrical.
        5. Lower surface of fronds flattened or only weakly convex, apex rounded to acute, fronds usually 3-nerved_____4. *L. minor*

5. Lower surface of fronds moderately to strongly convex, apex broadly rounded or often obtuse; fronds 3- or 5-nerved_____2. *L. gibba*

4. Rootlet sheaths winged.

  6. Fronds with a single apical papilla; seeds with 8–26 strong ribs_____
    _____1. *L. aequinoctialis*

  6. Fronds with 2–3 papillae; seeds with 35 or more indistinct ribs.

    7. Fronds firm, weakly to strongly asymmetrical, the main nerves often inconspicuous_____6. *L. perpusilla*

    7. Fronds membranous, symmetrical or nearly so, the main nerves prominent_____7. *L. trisulca*

3. Fronds distinctly 1-nerved or obscurely nerved.

  8. Fronds 1-nerved, the lower surface green.

    9. Fronds ovate-elliptic, weakly to moderately asymmetrical, often floating in compact masses_____9. *L. valdiviana*

    9. Fronds ovate to orbicular, symmetrical, usually solitary_____
    _____3. *L. minuta*

  8. Fronds obscurely nerved, the lower surface frequently reddish purple_
    _____5. *L. obscura*

Page 142. *Lemna trisulca* L. Add the following counties to the map at the top of page 143: DeKalb, Fulton, Greene, Jackson, McHenry, Rock Island, Will.

Page 143. *Lemna minor* L. Add the following counties to the map at the bottom of page 143: Calhoun, Champaign, Crawford, Cumberland, DeWitt, Effingham, Fayette, Gallatin, Jasper, Johnson, Lawrence, Logan, McDonough, Piatt, Saline, Schuyler, Wabash, Williamson.

Page 145. *Lemna gibba* L. There are no new records for this species.

Page 146. *Lemna perpusilla* Torr. Add the following counties to the map on page 146: Alexander, Bureau, Cook, Effingham, JoDaviess, Lake, Whiteside.

Page 146. *Lemna trinervis* (Austin) Small. There are no new records for this species.

Page 148. *Lemna valdiviana* Phil. Add the following county to the map at the bottom of page 148: Pope.

Page 149. *Lemna minima* Phil. is now known as **Lemna minuta** Kunth in HBK. Its nomenclature follows:

Lemna minuta Kunth in HBK. Nov. Gen. & Sp. 1:372. 1816.
Lemna minima Phil. Linnaea 33:239. 1864.

Lemna valdiviana Phil. var. *minima* (Phil.) Hegelm. Lemnac. 138. 1868.

There are no new records for this species.

Page 149. *Lemna obscura* (Austin) Daubs. Add the following county to the map on page 150: Jackson.

Page 150. Following *Lemna obscura* (Austin) Daubs, add the following species to the Illinois flora:

Lemna aequinoctialis Welw. Apont. Phyto. 578. 1858. Fig. A7.

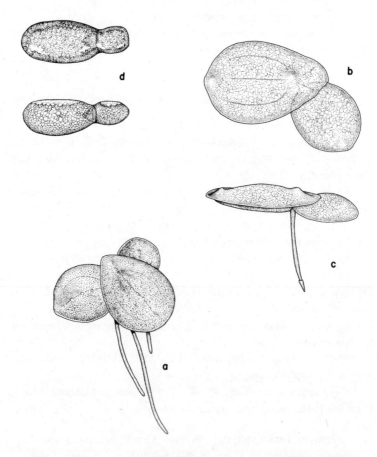

A7. *Lemna turionifera* (Turion-forming Duckweed). a. Habit. *Lemna aequinoctialis*. b. Top view. c. Side view. *Wolffia globosa* (Water Meal). d. Fronds.

Fronds single or connected in clusters of 2 or 3, ovate to nearly orbicular, symmetrical, firm, flattened, 3-nerved, with a single apical papilla; rootlet 1, the sheath narrowly winged, the tip acute; fruit 0.5–0.8 mm long, with 8–26 strong ribs.

COMMON NAME: Duckweed.
HABITAT: Standing water.
RANGE: Virginia to Nebraska, south to California, Texas, and Florida.
ILLINOIS DISTRIBUTION: Scattered in southern Illinois; Jackson, Union, and Williamson counties.

This species differs by the single apical papilla on the fronds.

Lemna turionifera Landolt, Aquat. Bot. 1:355. 1975. Fig. A7.

Fronds single to connected in clusters of 2–3, ovate to nearly orbicular, flattened to slightly convex below, sometimes with reddish margins, 1.5–4.0 mm long, with 2–4 papillae, producing rounded, rootless, olive to brown turions late in the season; rootlet 1, the sheath unwinged, the tip obtuse; fruit 1.0–1.4 mm long; seeds with 30–60 indistinct ribs.

COMMON NAME: Turion-forming Duckweed.
HABITAT: Standing water.
RANGE: Prince Edward Island to Alaska, south to California, Texas, northern Alabama, and Virginia.
ILLINOIS DISTRIBUTION: Scattered in the state: Clinton, Shelby, Union, and Williamson counties.

This species produces turions late in the season, thus distinguishing it from other species of *Lemna* in Illinois. Turions are specialized, overwintering fronds.

Page 150. *Wolffiella floridana* (J. D. Smith) Thompson. The correct binomial for this species appears to be **Wolffiella gladiata** (Hegelm.) Hegelm. Its taxonomy follows:

Wolffiella gladiata (Hegelm.) Hegelm. Bot. Jahrb. Syst. 21:304–305. 1895.
Wolffia gladiata Hegelm. Die Lemnaceean 133. 1868.
Wolffiella gladiata (Hegelm.) Hegelm. var. *floridana* J. D. Smith, Bull. Torrey Club 7:64. 1880.
Wolffiella floridana (J. D. Smith) Thompson, Rep. Mo. Bot. Gard. 9:37. 1898.

The following county should be added to the map on page 152: Williamson.

Page 152. *Wolffia* Horkel. A fourth species of *Wolffia* has been added to the Illinois flora. A new key to the genus in Illinois is below:

1. Plants about as deep as wide, boat-shaped.
    2. Plants 1 ½–2 times longer than wide, subacute at apex, without a papule__ _____1. *W. borealis*
    2. Plants 1–1 ½ times longer than wide, obtuse at apex, with a prominent central papule_____2. *W. brasiliensis*
1. Plants deeper than wide, globose to ovoid.
    3. Plants up to 1.3 times longer than wide, 0.4–1.2 mm wide_____ _____3. *W. columbiana*
    3. Plants 1.3–2.0 times longer than wide, 0.3–0.5 mm wide_____ _____4. *W. globosa*

Page 152. *Wolffia papulifera* Thompson. The correct binomial for this species is **Wolffia brasiliensis** Weddell. Its taxonomy follows:

Wolffia brasiliensis Wedd. Ann. Sci. Nat. Bot. ser. 3, 12:170. 1849.
Wolffia papulifera Thompson, Rep. Mo. Bot. Gard. 9:40. 1898.

Add the following counties to the map at the top of page 153: Hancock, Williamson.

Page 153. *Wolffia punctata* Griseb. This species is now called **Wolffia borealis** (Engelm.) Landolt. Its taxonomy follows:

Wolffia borealis (Engelm.) Landolt, Ber. Geobot. Inst. Eidgen. Tech. Hoch. 44:137. 1977.
Wolffia punctata Griseb. Fl. Brit. W. Ind. 512. 1864, misapplied.
Wolffia brasiliensis Wedd. var. *borealis* Engelm. Die Lemnaceen 127. 1868.

Add the following counties to the map at the bottom of page 153: Adams, DuPage, Hancock, Jackson, Richland.

Page 153. *Wolffia columbiana* Karst. Add the following counties to the map on page 154: Boone, Calhoun, Cass, Clark, Clay, Coles, Crawford, Cumberland, DeKalb, DuPage, Douglas, Edgar, Greene, Hancock, Jackson, Jasper, Kendall, Lawrence, Logan, McHenry, Morgan, Pope, Wabash, Williamson, Woodford.

Page 154. After *Wolffia columbiana* Karst, add the following new species of *Wolffia* for Illinois:

Wolffia globosa (Roxb.) Hartog & Plas, Blumea 18:367. 1970.
   Fig. A7.
Lemna globosa Roxb. Fl. Ind. 3:565. 1832.

Fronds ovoid, 0.4–0.8 mm long, 0.3–0.5 mm wide, up to twice as long as broad, 1–1 ½ times as deep as broad, rounded or slightly pointed at the tip; papilla 0; pigmented cells absent in vegetative tissue; rootlets absent; fronds solitary but sometimes remaining attached in groups; flowers and fruits rarely observed.

COMMON NAME: Water Meal.
HABITAT: Standing water.
RANGE: New Hampshire to Manitoba, south to California, Texas, and Florida.
ILLINOIS DISTRIBUTION: Scattered throughout the state.

Page 156. *Sparganium minimum* Fries. The correct binomial for this species is **Sparganium natans** L. Its taxonomy follows:

Sparganium natans L. Sp. Pl. 1:971. 1753.
Sparganium minimum Fries, Summa Veg. Scand. 2:560. 1849.

There are no new records for this species in Illinois.

Page 156. *Sparganium chlorocarpum* Rydb. The correct binomial for this species appears to be **Sparganium emersum** Rehmann. Its taxonomy follows:

Sparganium emersum Rehm. Verh. Naturf. Vereins Brunn 10:80. 1872.
Sparganium simplex Hudson var. *acaule* Beeby ex Macoun, Cat. Canad. Pl. 5:367. 1909.
Sparganium acaule (Beeby) Rydb. N. Am. Fl. 17:8. 1909.
Sparganium chlorocarpum Rydb. N. Am. Fl. 17:8. 1909.
Sparganium chlorocarpum Rydb. var. *acaule* (Beeby) Fern. Rhodora 24:29. 1922.

Add the following counties to the map on page 158: Iroquois, Kane, McHenry.

Page 158. *Sparganium androcladum* (Engelm.) Morong. Add the following county to the map on the top of page 161: St. Clair.

Page 161. *Sparganium americanum* Nutt. Add the following counties to the map at the bottom of page 161: Fulton, Kane, Knox, Pike, Stephenson.

Page 163. *Sparganium eurycarpum* Engelm. Add the following counties to the map on page 163: Crawford, Fayette, Jackson, Monroe.

Page 164. *Typha* L. A new key for *Typha* in Illinois follows. It is modified from the *Typha* key by Galen Smith in Flora North America (2003).

1. Pistillate bracteoles either absent or narrower than the stigmas, not evident at the spike surface; pistillate spikes green in flower when fresh, 19–36 mm thick in fruit.

  2. Pistillate bracteoles absent; pistillate spikes contiguous usually with staminate spikes, in fruit 24–36 mm thick; seeds numerous; pollen in tetrads___ _____3. *T. latifolia*

  2. Pistillate bracteoles present; pistillate spikes usually separated from staminate spikes by a gap, in fruit 19–25 mm thick; seeds few or none; pollen a mixture of 1–4 grains.

    3. Mucilage glands absent from blade; pistillate spikes after flowering medium to dark brown_____*T. angustifolia* X *T. latifolia*

    3. Mucilage glands usually present on adaxial surface of blade near the sheath; pistillate spikes after flowering bright orange..*T. domingensis* X *T. latifolia**

1. Pistillate bracteoles as wide as or wider than the stigmas, evident at the spike surface; pistillate spikes brown at all stages, or white when fresh, 13–25 mm thick in fruit.

  4. Mucilage glands absent from surface of blade; pistillate bracteoles darker than stigmas; pistillate spikes medium to dark brown_____ _____1. *T. angustifolia*

  4. Mucilage glands present on adaxial surface of all of sheath and usually part of adjacent blade; pistillate bracteoles paler or same color as stigmas; pistillate spikes bright cinnamon- to orange-brown to medium brown.

    5. Pistillate bracteoles paler than the stigmas; pistillate spikes usually bright cinnamon- to orange-brown; mucilage glands numerous on leaf blade__ _____*T. domingensis*

    5. Pistillate bracteoles the same color as the stigmas; pistillate spikes medium brown; mucilage glands few or absent from leaf blade_____ _____*T. angustifolia* X *T. domingensis**

Page 164. *Typha latifolia* L. There are no new records for this species.
Page 166. *Typha angustifolia* L. Add the following counties to the map on page 166: Brown, Carroll, Clark, Coles, Cumberland, DeWitt, Edgar, Gallatin, Hamilton, Hancock, Iroquois, Jefferson, JoDaviess,

Lawrence, Lee, Livingston, Macon, Madison, Marion, McHenry, Menard, Ogle, Rock Island, Schuyler, Shelby, Stephenson, Whiteside.

After *T. angustifolia* L., add the following to the Illinois flora:

Typha domingensis Persoon, Syn. Pl. 2:532. 1807. Fig. A8.

Coarse perennial from stout, creeping rhizomes; stems to 4 m tall, with numerous cauline leaves; leaves elongated, more or less convex, to 18 mm wide, with mucilage glands at the sheath-blade transition and along the entire sheath and part of the blade; staminate spikes separated from the pistillate spikes by up to 8 cm, to 35 cm long, to 1 cm thick at anthesis, the scales stramineous to bright orange-brown, linear, with pollen grains borne singly; pistillate spikes bright cinnamon-brown when fresh, becoming bright orange-brown, to 35 cm long, to 6 mm wide at anthesis, to 25 mm wide in fruit, the bracteoles paler than or about the same color as the stigmas; achenes about 1 mm long, surrounded by numerous whitish hairs.

COMMON NAME: Southern Cat-tail.
HABITAT: Known originally from a power plant cooling pond in Illinois.
RANGE: Virginia to Illinois to Wyoming, south to California, Texas, and Florida.
ILLINOIS DISTRIBUTION: Jackson and Lake counties.

This species seems to have the broad leaves of *T. latifolia* but the narrower pistillate spikes of *T. angustifolia*. Its height is usually considerably greater than in the other two species of *Typha*. *Typha domingensis* flowers from June to September.

Typha domingensis hybridizes with *T. latifolia* to form *T. X provincialis* A. Camus, and with *T. angustifolia*. The distinguishing characteristics of the hybrids may be found in the key above. The hybrid with *T. latifolia* is known from Missouri and Nebraska.

The hybrid between *T. latifolia* L. and *T. angustifolia* L., called **T. X glauca** Godr., has been found in wet ground in DuPage, Kane, and Rock Island counties.

Page 169. *Xyris torta* L. Add the following county to the map on page 169: Menard.

Page 169. *Xyris jupicai* L. There are no new records for this species.

Page 173. *Tradescantia subaspera* Ker. Two varieties of this species now occur in Illinois, separated by the following key:

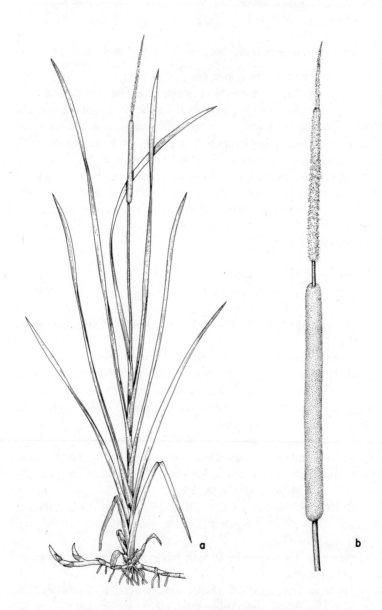

A8. Typha domingensis (Southern Cat-tail). a. Habit. b. Flowering spike.

1. Some or all of the uppermost lateral cymes sessile_____

_____*T. subaspera* var. *subaspera*

1. All of the lateral cymes pedunculate_____*T. subaspera* var. *montana*

Page 173. *Tradescantia subaspera* Ker var. *subaspera*. Add the following counties to the map on page 173: Cumberland, Douglas, Effingham, Fayette, Fulton, Greene, Morgan, Montgomery, Scott, Shelby, St. Clair, Washington.

Tradescantia subaspera Ker var. **montana** (Shuttlew.) Anderson & Woodson, Contr. Arn. Arb. 9:52. 1935. Not illustrated.

Tradescantia montana Shuttlew. ex Small & Vail, Mem. Torrey Club 4:150. 1893.

Tradescantia subaspera Ker ssp. *montana* (Shuttlew.) R. T. Clausen, Cornell Univ. Ag. Exp. Sta. Mem. 291:10. 1949.

All of the lateral cymes pedunculate.

COMMON NAME: Mountain broad-leaved spiderwort.

HABITAT: Rich woods.

RANGE: North Carolina, Tennessee, Illinois.

ILLINOIS DISTRIBUTION: Jackson County (Fountain Bluff), collected by Keith Wilson in 1976.

This variety is usually a plant of the Appalachian Mountains. The station at Fountain Bluff in Jackson County is very disjunct.

Page 175. *Tradescantia ohiensis* Raf. Add the following counties to the map on page 175: Clark, Crawford, Effingham, Gallatin, Hardin, Marion, Montgomery, Saline, Shelby, Vermilion, Williamson.

Page 175. *Tradescantia virginiana* L. Add the following counties to the map on page 178: Fulton, Kane, Macon, Tazewell.

Page 178. *Tradescantia bracteata* Small. Add the following counties to the map on page 180: Adams, Greene, Hancock, Henry, Madison, McDonough, Menard, Peoria.

Page 180. *Commelina communis* L. Add the following counties to the map at the top of page 182: Clark, Cumberland, Effingham, Franklin, Gallatin, Hamilton, Jefferson, Knox, Macoupin, Montgomery, Perry, Randolph, Saline, Schuyler, White.

Page 182: *Commelina diffusa* Burm. f. Add the following counties to the map at the bottom of page 182: Calhoun, Christian, Macon, Menard, Sangamon.

Page 184. *Commelina virginica* L. Add the following counties to the map on page 184: Crawford, Lawrence, McDonough, Williamson.

Page 184. *Commelina erecta* L. Three varieties may be recognized in Illinois, distinguished by the following key:

1. Stems 0.6–1.2 m tall; leaves lanceolate to lance-ovate, some or all of them 2–4 cm wide_____*C. erecta* var. *erecta*
1. Stems up to 0.6 m tall; leaves linear to linear-lanceolate, up to 2 cm wide.
    2. Spathes 1.0–2.5 cm long; leaves 4–10 cm long_____
    _____*C. erecta* var. *angustifolia*
    2. Spathes 2.5–3.0 cm long; leaves 7–15 cm long____*C. erecta* var. *deamii*

Commelina erecta L. var. **erecta**

Add the following county to the map on page 187: Iroquois. Delete the following counties from the map on page 187: Cook, Kankakee.

Commelina erecta L. var. **angustifolia** (Michx.) Fern. Rhodora 42:439. 1940. Not illustrated.
Commelina angustifolia Michx. Fl. Bor. Am. 1:24. 1803.
Commelina crispa Woot. Bull. Torrey Club 25:451. 1898.
Commelina erecta L. var. *angustifolia* (Michx.) Fern. f. *crispa* (Woot.) Fern. Rhodora 42:440. 1940.

Stems up to 0.6 m tall; leaves linear to linear-lanceolate, up to 10 cm long, up to 2 cm wide; spathes 1.0–2.5 cm long.

COMMON NAME: Narrow-leaved day-flower.
HABITAT: Dry, sandy soil.
RANGE: Similar to var. *erecta.*
ILLINOIS DISTRIBUTION: Alexander, Grundy, Jackson, Massac, Pope, Pulaski, Union.
This variety flowers from June to September.

Commelina erecta L. var. **deamiana** Fern. Rhodora 42:449. 1940. Not illustrated.

Stems up to 0.6 m tall; leaves linear, 7–15 cm long, up to 2 cm wide; spathes 2.5–3.0 cm long.

COMMON NAME: Deam's day-flower.
HABITAT: Dry, sandy soil.

RANGE: Similar to var. *erecta*.

ILLINOIS DISTRIBUTION: Cook and Kankakee counties.

This variety flowers from June to September.

Page 187. Pontederiaceae. A new key to the genera of Pontederiaceae in Illinois is provided below since *Eichhornia* has now been found in the state:

KEY TO THE GENERA OF Pontederiaceae IN ILLINOIS

1. Inflorescence with 50 or more flowers; fruit a 1-seeded utricle_____
_____*Pontederia*
1. Inflorescence with 1–30 flowers; fruit a capsule with 10–many seeds.
  2. Leaves linear; inflorescence 1-flowered_____*Zosterella*
  2. Leaves lanceolate to ovate to reniform; inflorescence 1- to several-flowered.
    3. Petiole swollen; perianth lobes more than 2 cm long; stamens 6_____
    _____*Eichhornia*
    3. Petiole not swollen; perianth lobes less than 2 cm long; stamens 3_____
    _____*Heteranthera*

Page 188. *Pontederia cordata* L. The following counties should be added to the map on page 188: DeKalb, Gallatin.

Page 190. *Zosterella dubia* (Jacq.) Small. The following counties should be added to the map on page 190: Boone, Bureau, DeKalb, Edgar, Fulton, Grundy, Hancock, Jersey, JoDaviess, Kane, Kankakee, Macoupin, Madison, Morgan, Ogle, Rock Island, Vermilion, Wabash, Will.

Page 192. *Heteranthera* R. & P. Two additional species have been discovered in Illinois. The following key will serve to distinguish them:

KEY TO THE SPECIES OF Heteranthera IN ILLINOIS

1. Stems short, erect, with the leaves clustered at the tip; leaves truncate to cuneate_____1. *H. limosa*
1. Stems creeping, with scattered leaves; leaves cordate.
  2. Flower solitary; leaves longer than broad_____4. *H. rotundifolia*
  2. Flowers 1–3 in a cluster; leaves as broad as long or broader.
    3. Flowering spathes more or less sessile; perianth purple_____
    _____2. *H. multiflora*
    3. Flowering spathes on short stalks; perianth white_____
    _____3. *H. reniformis*

Page 192. *Heteranthera limosa* (Sw.) Willd. The following counties should be added to the map on page 193: Adams, Cass, Macon, Union.

Page 194. *Heteranthera reniformis* R. & P. There are no new records for this species.

After *Heteranthera reniformis* R & P., add the following:

Heteranthera multiflora (Griseb.) C. N. Horn, Phytologia 59 (4):290. 1986. Fig. A9.

A9. *Heteranthera multiflora*
(Mud Plantain). Habit.

Heteranthera reniformis Ruiz & Pav. var. *multiflora* Griseb. Abh. Kon. Ges. Wiss. Gott. 24:323. 1879.

Perennial from rhizomes; leaves orbicular, subacute at the apex, cordate at the base, to 5 cm long, nearly as broad, with a petiole to 15 cm long; inflorescence spicate, 2- to 8-flowered, the flowers purple; spathe sessile, short-acuminate, 1–3 cm long; perianth tube 5–10 cm long, the inner segments linear-lanceolate, the outer segments narrower; filaments pilose; stigma capitate; capsule oblongoid, 5–9 mm long.

COMMON NAME: Mud Plantain.
HABITAT: Muddy shores; shallow water.
RANGE: Delaware, Maryland, New Jersey, Pennsylvania; Illinois to Iowa to Nebraska, south to Texas and Mississippi; South America.
ILLINOIS DISTRIBUTION: Hardin County.

This species differs from the very similar *H. reniformis* by its purple flowers and its sessile spathes. It flowers during July and August.

Heteranthera rotundifolia (Kunth) Griseb. Cat. Pl. Cub. 252–53. 1866. Fig. A10.

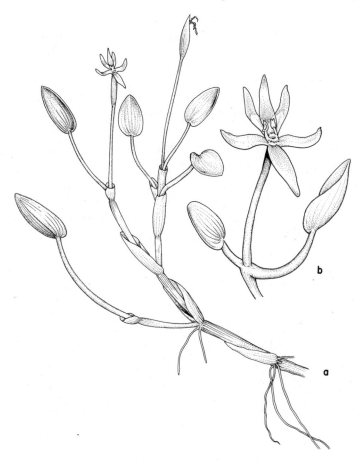

A10. Heteranthera rotundifolia (Mud Plantain). a. Habit. b. Flower, with upper leaves.

Heteranthera limosa (Sw.) Willd. var. *rotundifolia* Kunth, Enum. Pl. Omn. Huc. Cog. 4:122. 1843.

Annual with creeping stems; leaves dimorphic: rosette leaves linear, emergent leaves to 5 cm long, lanceolate to ovate, obtuse at apex, truncate or subcordate at base, long-petiolate; flower borne singly, purple to white, 2-lipped, the lobes linear, to 18 mm long, the lateral lobes spreading, the middle lobe descending with a yellow spot at base, the tube to 30 mm long; stamens unequal, the filaments glandular-hairy.

COMMON NAME: Mud Plantain.

HABITAT: Muddy shores, shallow water.

RANGE: Kentucky to Nebraska and Colorado, south to Arizona and Texas; California; West Indies; Central America; South America.

ILLINOIS DISTRIBUTION: Alexander County.

This species is distinguished by its solitary flower and its leaves longer than broad. This species flowers during July and August.

Eichhornia Kunth—Water Hyacinth

Perennials or annuals; leaves with parallel veins, petiolate, usually with a swollen area on the petioles; stipules absent; inflorescence spicate or a spikelike panicle; perianth 6-parted, united below the middle, more or less 2-lipped; stamens 6; fruits capsules enclosed by the persistent perianth; seeds numerous, ribbed.

There are eight tropical and subtropical species in the genus.

Eichhornia crassipes (Mart.) Solms, Monogr. Phan. 4:527. 1883. Fig. A11.

Pontederia crassipes Mart. Nova Gen. & Sp. Pl. 1:9, pl. 4. 1823 (1824).

Annuals (in our area); stems stout; leaves reniform to orbicular, shiny, 3–10 cm wide, on petioles up to 20 (–35) cm long, bulbous-inflated at the middle; inflorescence a spicate panicle, glandular-pubescent; perianth more or less 2-lipped, the lobes 3–4 cm long, usually lilac, the upper lobe yellow at base; stamens 6, 3 long and 3 short; capsules to 2 cm long; seeds 1 mm long, longitudinally ribbed.

COMMON NAME: Water Hyacinth.

HABITAT: Shallow water.

RANGE: Native to South America; commonly adventive in the southeastern United States.

ILLINOIS DISTRIBUTION: Massac County.

This wetland adventive is a nuisance plant in the southeastern United States but barely survives in Illinois.

Page 196. Since I now believe *Luzula echinata* and *L. bulbosa* to be distinct species, I am providing a new key to the species of *Luzula* in Illinois:

A11. Eichhornia crassipes (Water Hyacinth). Habit.

KEY TO THE SPECIES OF Luzula IN ILLINOIS

1. Flower solitary (rarely paired) at the tips of the inflorescence rays; seeds more than 2 mm long, including the strongly curved caruncle_____*L. acuminata*
1. Flowers crowded in glomerulate spikes; seeds 1.2–2.0 mm long, including the conical caruncle.
    2. Plants producing small, white bulblets at base_____*L. bulbosa*
    2. Plants not producing small, white bulblets at base.
        3. Rays of umbel erect to ascending_____*L. multiflora*
        3. Rays of umbel horizontally spreading to reflexed_____*L. echinata*

Page 196. *Luzula acuminata* Raf. There are no new records for this species.

Page 198. *Luzula multiflora* (Retz.) Lejeune. Because I now recognize *L. bulbosa* as distinct and not synonymous with *L. multiflora,* and since I now believe that *L. echinata* should be recognized at the species level, rather than as a variety of *L. multiflora,* new taxonomy is provided for these three species:

Luzula multiflora (Retz.) Lejeune, Fl. Envir. Spa. 1:169. 1811.
Juncus multiflora Retz. Fl. Scand. Prodr. 82. 1795.
Luzula campestris (L.) DC. var. *multiflora* (Retz.) Celak, Prodr. Fl. Boehem. 85. 1869.

Add the following counties to the map on page 198: Iroquois, Williamson.

Luzula echinata (Small) Hermann, Rhodora 40:84. 1938.
Juncoides echinatum Small, Torreya 1:74. 1901.
Luzula campestris (L.) DC. var. *echinata* (Small) Fern. & Wieg. Rhodora 15:42. 1913.
Luzula echinata (Small) Hermann var. *mesochorea* Hermann, Rhodora 40:84. 1938.
Luzula multiflora (Retz.) Lejeune var. *echinata* (Small) Mohlenbr., Ill. Fl. Il. Flowering Rush to Rushes 200. 1970.

The following county should be added to the map on page 200: Williamson.

Luzula bulbosa (Wood) Rydb. Brittonia 1:85. 1931. Fig. A12.
Luzula campestris (L.) DC. var. *bulbosa* Wood, Class-Book 723. 1861.

Perennial with short rhizomes bearing white bulblike tubers; stems to 40 cm tall; leaves lance-linear, to 7 mm broad, ciliate; inflorescence umbellate, the central spike often sessile, the 1–20 other spikes on erect rays; spikes 4–12 mm long, 5–7 mm broad; perianth segments lanceolate, acute to acuminate, 2–3 mm long, chestnut with hyaline margins and tip;

A12. Luzula bulbosa (Bulbous Wood Rush). a. Habit. b. Leaf and part of stem. c. Capsule, with perianth, side view. d. Capsule, top view.

capsules obovoid, brown, 2.5–4.0 mm long; seeds dark brown, 0.9–1.3 mm long, with a conical caruncle to 0.7 mm long.

COMMON NAME: Bulbous Wood Rush.

HABITAT: Woods, rocky areas.

RANGE: Delaware to Illinois, south to east Texas and north Florida.

ILLINOIS DISTRIBUTION: Confined to southern Illinois: Gallatin, Jackson, Johnson, Pope, Saline, Union, and Williamson counties.

This species is readily distinguished from the other species of *Luzula* in Illinois by the presence of white bulblets at the base of the plant.

Page 200. *Juncus* L. Several additional species of *Juncus* have been found in Illinois since the publication of the first edition of Flowering Rush to Rushes (1970). Both technical and nontechnical keys to the species of *Juncus* in Illinois are below:

TECHNICAL KEY

1. Leaf sheaths without blades, apiculate or mucronate; inflorescence lateral on the stems.
    2. Stems densely cespitose; stamens 3; anthers 0.5–0.8 mm long; seeds 0.5 mm long; capsules beakless_____*J. effusus*
    2. Stems single at intervals from elongated rhizomes; stamens 6; anthers 1.5–2.0 mm long; seeds 1 mm long; capsules with beak 0.5–1.0 mm long_____
    _____*J. arcticus*
1. Leaf sheaths with definite blades; inflorescence terminal.
    3. Flowers in heads, not prophyllate (i.e., not with bracteoles).
        4. Leaves flat, not terete or cross-septate; anthers purplish brown.
            5. Stems solitary, approximately 33.5 cm apart on conspicuous, scaly rhizomes, 4.9–8.8 (–10.2) dm tall; leaves 2.0–6.5 mm wide; heads (13–) 20–135_____*J. biflorus*
            5. Stems cespitose, 0.45–0.90 dm tall; leaves 1.0–2.5 (–2.9) mm wide; heads 3–28 (–32)_____*J. marginatus*
        4. Leaves terete and hollow (flat and hollow in *J. validus*) and usually cross-septate; anthers yellow.
            6. Seeds fusiform, 0.7–1.9 mm long, caudate.
                7. Seeds 1.2–1.9 mm long, the tails comprising (⅓–) ½–⅝ the total length of the seeds; heads 5- to 50-flowered; stamens 3_____
                _____*J. canadensis*
                7. Seeds 0.7–1.2 mm long, the tails comprising ¼–⅔ the total length of the seeds; heads 2- to 5- (to 10-) flowered; stamens 3 or 6.

8. Sepals obtuse to subacute_____*J. brachycephalus*

8. Sepals acute to acuminate_____*J. subcaudatus*

6. Seeds ellipsoid to oblongoid to ovoid, 0.4–0.6 mm long, apiculate.

9. Stamens 6.

10. Involucral leaves shorter than the inflorescences; heads hemi-spherical or ellipsoid, 2–7 mm wide, 2- to 9-flowered; sepals 1.9–3.0 mm long, acute to acuminate to obtuse; capsules oblon-goid or ellipsoid, acute to obtuse.

11. Petals mostly shorter and blunter than the sepals; capsules ob-tuse and abruptly short-pointed_____*J. alpinoarticulatus*

11. Petals as long as or a little longer than the sepals; capsules more attenuate at tip_____*J. articulatus*

10. Involucral leaves usually exceeding the inflorescences; heads spherical or hemispherical, 8–15 mm wide, 9- to 90-flowered; se-pals 2.5–5.0 mm long, subulate; capsules lanceoloid, subulate.

12. Sepals 2.5–4.0 mm long; petals equaling to exceeding the sepals by 0.8 mm; heads 8–11 (–12) mm across; stems to 6 dm tall_____*J. nodosus*

12. Sepals 4–5 mm long; petals 1 mm shorter than to nearly equal-ing the sepals; heads 10–15 mm across; stems to 10.7 dm tall_____*J. torreyi*

9. Stamens 3.

13. Capsules linear-lanceoloid to lanceoloid, acute, exceeding the sepals by more than 1 mm, at least 1.5 mm.

14. Capsules linear-lanceoloid, twice as long as the petals or lon-ger_____*J. diffusissimus*

14. Capsules lanceoloid, about ⅓ longer than the petals_____ _____*J. debilis*

13. Capsules ellipsoid or oblongoid, obtuse to subulate, shorter than to exceeding the sepals only by 1 mm.

15. Capsules oblongoid, subulate, exceeding the sepals by 0.75–1.00 mm; perianth segments subulate.

16. Leaves hollow and terete_____*J. scirpoides*

16. Leaves hollow and flattened_____*J. validus*

15. Capsules ellipsoid, acute to obtuse, shorter than to exceed-ing the sepals by 0.75 mm; perianth segments subulate to acuminate.

17. Perianth segments acuminate; heads 2- to 35-flowered, hemispherical to spherical; petals 0.7 mm shorter than

to equaling the sepals; capsules slightly shorter than the petals or exceeding them by 0.75 mm.

    18. Sepals 2.0–2.5 mm long; heads 150–280, 3–5 mm across, 2- to 7- (to 8-) flowered; leaves 1.5–4.5 mm wide_____*J. nodatus*

    18. Sepals 3–4 mm long; heads (1–) 2–82, 5–10 mm across, 5- to 35-flowered; leaves 1–3 mm wide_____
_____*J. acuminatus*

  17. Perianth segments subulate; heads densely 50- to 80-flowered, spherical; petals distinctly (approximately 1 mm) shorter than the sepals; capsules usually 0.5 mm shorter than the sepals_____*Juncus brachycarpus*

3. Flowers borne singly on the inflorescence branches, not in heads, prophyllate (i.e., with bracteoles).

  19. Annual; auricle at top of sheath absent; sepals 4–7 mm long; inflorescence comprising ¼–⅛ the total height of the plant_____*J. bufonius*

  19. Perennial; auricle at top of sheath present; sepals 2.3–6.0 mm long; inflorescence comprising less than ½ the total height of the plant.

    20. Sepals obtuse; anthers 1 mm long, 3 times longer than the filaments_
_____*J. gerardii*

    20. Sepals acuminate, subulate, or aristate; anthers shorter than to as long as the filaments.

      21. Leaves terete, as least distally; capsules usually exceeding the perianth; seeds 0.5–1.3 mm long.

        22. Leaves involute near the summit of the sheath, becoming closed and terete above; inflorescences 1.5–6.5 (–8.0) cm long; petals acute or obtuse; capsules exceeding the sepals by (0.75–) 1.0–1.6 mm; seeds 0.5–0.6 mm long, apiculate at both ends_____*J. greenei*

        22. Leaves terete throughout; inflorescences (1.0–) 2.0–3.5 cm long; petals acuminate or aristate; capsules slightly shorter than to exceeding the sepals by 1 mm; seeds 1.0–1.3 mm long, caudate at both ends_____*J. vaseyi*

      21. Leaves flat or involute; capsules shorter than to exceeding the perianth by 0.1 mm; seeds 0.3–0.5 mm long.

        23. Petals 0.2 mm shorter than to equaling the sepals; tips of the inflorescence branches incurved; leaves to 13 cm long_____
_____*J. secundus*

        23. Petals 1 mm shorter than to equaling the sepals; tips of the inflorescence branches not incurved; leaves to 30 cm long.

24. Auricles friable, not firm or rigid, scarious, hyaline, prolonged 1.0–4.5 (–5.0) mm beyond point of insertion; perianth segments spreading.

    25. Capsules no longer than the length of the sepals, the flowers more or less interrupted in the inflorescence; plants at least usually 70 cm tall____*J. anthelatus*

    25. Capsules no shorter than the length of the sepals, the flowers congested in the inflorescence; plants usually less than 70 cm tall'_____*J. tenuis*

24. Auricles firm at apex or rigid, cartilaginous or membranous, occasionally prolonged to 2 mm beyond point of insertion; perianth segments spreading or appressed.

    26. Auricles cartilaginous, opaque, rigid, often slightly flaring, obtuse, yellow or orange-brown, less than 1 mm long and usually 0.75 mm prolonged beyond point of insertion; perianth segments spreading; bracteoles obtuse to acute_____*J. dudleyi*

    26. Auricles membranous, hyaline, not cartilaginous or rigid, usually firm at apex, pale or brown, very slightly prolonged to exserted 2 mm beyond point of insertion; perianth segments appressed; bracteoles acuminate to aristate_____*J. interior*

NONTECHNICAL KEY

1. Leaves absent; flowers appearing laterally.

    2. Stems without longitudinal ribs; stems in rows from a rhizome_____ _____*J. arcticus*

    2. Stems with longitudinal ribs; stems cespitose_____*J. effusus*

1. Leaves present; flowers appearing terminally.

    3. Flowers prophyllate, subtended by 2 small opposite bracteoles.

        4. Annuals; inflorescence ⅓ or more the total height of the plant_____ _____*J. bufonius*

        4. Perennials; inflorescence less than ½ the total height of the plant.

            5. 2–4 leaves present on the stem_____*J. gerardii*

            5. Cauline leaves absent; all leaves basal.

                6. Leaves terete for some or all of their length.

                    7. Sepals 3.5–4.5 mm long; capsules 4.0–5.5 mm long_____ _____*J. vaseyi*

                    7. Sepals 2.3–3.5 mm long; capsules 3–4 mm long_____*J. greenei*

                6. Leaves flat or involute.

8. Flowers conspicuously on one side of branches_____*J. secundus*
8. Flowers not on one side of branches.
   9. Auricles scarious, projecting 1–3 mm beyond base of blade.
      10. Capsules no longer than the length of the sepals; flowers more or less interrupted in the inflorescence; plants usually at least 70 cm tall_____*J. anthelatus*
      10. Capsules no shorter than the length of the sepals; flowers congested in the inflorescence; plants usually less than 70 cm tall_____*J. tenuis*
   9. Auricles membranous or cartilaginous, not prolonged beyond base of blade.
      11. Auricles membranous, green or pale_____*J. interior*
      11. Auricles cartilaginous, yellow_____*J. dudleyi*
3. Flowers not prophyllate, not subtended by 2 small, opposite bracteoles.
  12. Leaves flat, not hollow.
     13. Plants with 15–20 heads, each head 6–8 mm across; blades 1–3 mm wide_____*J. marginatus*
     13. Plants with 20 or more heads, each head 4–6 mm across; blades 4–6 mm wide_____*J. biflorus*
  12. Leaves terete and hollow, or flat and hollow in *J. validus,* usually cross-septate.
     14. Leaves hollow and flat_____*J. validus*
     14. Leaves hollow and terete.
       15. Capsules at least twice as long as the perianth_____
       _____*J. diffusissimus*
       15. Capsules shorter than, equal to, or slightly longer than the perianth, but not twice as long as the perianth.
        16. Heads spherical or hemispherical.
         17. Heads hemispherical.
          18. Heads 6–10 mm in diameter.
           19. Seeds 0.3–0.4 mm long; flowers often more than 10 per head_____*J. acuminatus*
           19. Seeds 0.7–1.2 mm long; flowers usually 5 (–10) per head_____*J. subcaudatus*
          18. Heads 10–20 mm in diameter_____*J. canadensis*
         17. Heads spherical.
          20. Heads up to 10 mm in diameter.
           21. Sepals 3.5–6.0 mm long; heads 30–100 per plant_
           _____*J. brachycarpus*

21. Sepals 2.0–3.2 mm long; heads 20–60 per plant_
_____*J. scirpoides*
20. Heads 10–20 mm in diameter.
22. Leaves 0.5–1.5 mm wide; heads 5–25 per plant_
_____*J. nodosus*
22. Some of the leaves at least 2 mm wide; heads 25–
100 per plant_____*J. torreyi*
16. Heads wedge-shaped.
23. Heads 35–100 per plant.
24. Perianth about ⅔ as long as the capsules, obtuse to
acute_____*J. brachycephalus*
24. Perianth nearly as long as the capsules, acuminate__
_____*J. nodatus*
23. Heads 3–35 (–50) per plant.
25. Heads 3–5 mm across_____*J. debilis*
25. Heads 6–10 mm across.
26. Heads 6–8 mm across; capsules short-apiculate_
_____*J. alpinoaraticulatus*
26. Heads 8–10 mm across; capsules long-apiculate_
_____*J. articulatus*

Page 205. *Juncus effusus* L. var. *solutus* Fern. & Wieg. The following counties should be added to the map on page 205: Adams, Calhoun, Clark, Coles, Cumberland, DeKalb, Effingham, Gallatin, Kane, Lake, Ogle, Randolph.

Page 207. *Juncus balticus* Willd. var. *littoralis* Engelm. The binomial for this plant apparently should be **Juncus arcticus.** Its taxonomy follows:

Juncus arcticus Willd. Sp. Pl. 2 (1):206. 1799.
Juncus balticus Willd. var. *littoralis* Engelm. Trans. Acad. St. Louis 2:442. 1866.
Juncus arcticus Willd. var. *balticus* (Willd.) Trautv. Tgrudy Imp. St.-Petersburg Bot. Sada 5:119. 1875.
Juncus balticus Willd. var. *littoralis* Engelm. f. *dissitiflorus* Engelm. ex Fern. & Wieg. Rhodora 25:208. 1923.
Juncus litorum Rydb. Brittonia 1:85. 1931.

Add the following counties to the map on page 207: DuPage, LaSalle.

Page 209. *Juncus biflorus* Ell. Add the following counties to the map on page 209: Cass, Christian, Cook, Lee, Macoupin, Mason, Menard, Morgan, Randolph, Winnebago.

Page 211. *Juncus marginatus* Rostk. Add the following counties to the map on page 211: Alexander, Crawford, Edwards, Grundy, Jackson, Union, Wabash.

Page 213. *Juncus canadensis* J. Gay. Add the following counties to the map on page 213: Grundy, Jackson, Monroe, Pulaski, Richland, Union.

Page 215. *Juncus brachycephalus* (Engelm.) Buch. Add the following counties to the map on page 215: DuPage, Fayette, Jackson, Kendall, Macon, Pope, Will, Winnebago.

Page 217. *Juncus alpinus* Vill. This species is now known as **Juncus alpinoarticulatus** Chaix, and there are two subspecies. Their taxonomy follows:

Juncus alpinoarticulatus Chaix, Hist. Pl. Dauphine 1:378. 1786.
ssp. **alpinoarticulatus**
Juncus alpinus Vill. Hist. Pl. Dauphine 2:233. 1787.
Juncus alpinus Vill. var. *rariflorus* Hartm. Skand. Fl., ed. 7:140. 1868.

Juncus alpinoarticulatus Chaix ssp. **fuscescens** (Fern.) Hamat-Ahti, Bot. Fenn. 23. 280. 1986.
Juncus alpinus Vill. var. *fuscescens* Fern. Rhodora 10:48. 1908.

Add the following counties for ssp. *fuscescens* to the map on page 217: DuPage, Ogle.

Page 219. *Juncus nodosus* L. Add the following counties to the map on page 221: Boone, Christian, Coles, DeKalb, DeWitt, Fulton, Iroquois, Kendall, Shelby, St. Clair, Vermilion, Winnebago.

Page 221. *Juncus torreyi* Coville. Add the following counties to the map on page 223: Bond, Clay, Hamilton, Jackson, Pope, Union, Washington, White.

Page 224: *Juncus diffusissimus* Buckl. Add the following county to the map at the top of page 226: Pope.

Page 226. *Juncus scirpoides* Lam. Add the following county to the map at the bottom of page 236: Wabash.

Page 226. *Juncus nodatus* Coville. Add the following counties to the map on page 229: Alexander, Clark, Hamilton, Wabash.

Page 229. *Juncus acuminatus* Michx. Add the following counties to the map on page 231: Clark, Coles, DuPage, Gallatin, Iroquois, Kane, LaSalle, McDonough, Schuyler.

Page 231. *Juncus brachycarpus* Engelm. Add the following counties to the map on page 233: Cass, Clark, Coles, Gallatin, Mason, Wayne.

Page 233. *Juncus bufonius* L. Add the following counties to the map on page 235: Cass, DeKalb, Hardin, Iroquois, Jackson, Johnson, Kane, Kankakee, Menard, Monroe, Ogle, Randolph, St. Clair.

Page 235. *Juncus gerardii* Loisel. There are no new records for this species.

Page 237. *Juncus greenei* Oakes. The following counties should be added to the bottom of page 237: Fulton, Grundy, Ogle.

Page 238. *Juncus vaseyi* Engelm. There are no new records for this species.

Page 239. *Juncus secundus* Beauv. The following counties should be added to the map on page 242: Hancock, Jackson, Williamson.

Page 242. *Juncus tenuis* Willd. Add the following counties to the map on page 244: Ford, Moultrie, Stephenson.

Page 244. *Juncus dudleyi* Wieg. Add the following counties to the map on page 246: Bond, Brown, Christian, Clark, Effingham, Jefferson, Jersey, Lawrence, McLean, Perry, Pulaski, Randolph, Richland, Rock Island, Saline, Union, White.

Page 246. *Juncus interior* Wieg. Add the following counties to the map on page 248: Edgar, Greene, Grundy, JoDaviess.

After *Juncus interior* Wieg., add the following species that are new to Illinois:

Juncus anthelatus (Wieg.) R. E. Brooks, Novon 9 (1):11. 1999.
Fig. A13.
Juncus tenuis Willd. var. *anthelatus* Wieg. Bull. Torrey Club 27 (10):523–24. 1911.

Cespitose perennial; stems to 90 cm tall; leaves basal, flat to involute, to 30 cm long, 0.5–2.3 mm wide; inflorescence 10- to 100-flowered, with a bract exceeding the inflorescence; bracteoles 2; perianth parts green, lanceolate, 3.2–4.5 mm long, up to 1 mm wide, the outer and inner parts nearly equal; stamens 6; capsules tan, ellipsoid to obovoid, 2.0–3.2 mm long, 1.0–1.6 mm wide; seeds tan, ellipsoid, 0.3–0.6 mm long, not caudate.

COMMON NAME: Tall Path Rush.

HABITAT: Moist or wet soil, often in disturbed areas.

RANGE: New Brunswick to Ontario, south to Minnesota, northeast Texas, northern Georgia, and Virginia.

ILLINOIS DISTRIBUTION: Scattered throughout the state.

This species closely resembles *J. tenuis* except that its stems are almost always at least 70 cm tall. It flowers from May to September.

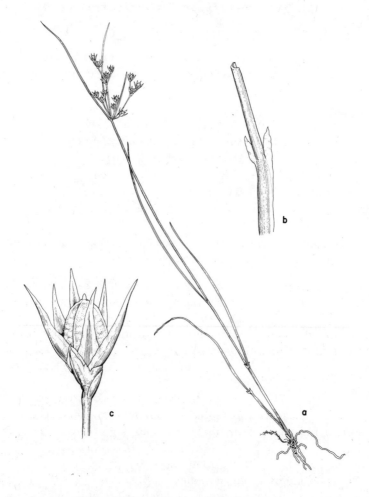

A13. Juncus anthelatus (Tall Path Rush). a. Habit. b. Auricles at top of sheath. c. Capsule, with perianth.

Juncus articulatus L. Sp. Pl. 1:327. 1753. Fig. A14.

Perennials from course rhizomes; stems tufted, up to 60 cm tall; leaves 2–4, up to 15 cm long, 0.7–1.5 mm in diameter, terete, hollow, cross-septate; involucral leaf usually shorter than the inflorescence; inflorescence up to 30 cm long, nearly twice as long as wide, with several hemispheric clusters, each cluster 8–10 mm across, with 3–10 flowers; flowers sessile or subsessile; perianth segments lanceolate to lance-subulate; sepals 2.0–2.7 mm long, acute to acuminate; petals 2.2–3.0 mm long, obtuse to subacute; stamens 6; capsules chestnut-brown, narrowly ovoid, acute, 2.5–4.0 mm long; seeds ovoid to oblongoid, acute or acuminate, 0.5 mm long, short-apiculate.

COMMON NAME: Jointed Rush.

HABITAT: Along a railroad in Illinois.

RANGE: Newfoundland to Alaska, south to Oregon, Arizona, northern Illinois, Kentucky, and Virginia; also in Europe and Asia.

ILLINOIS DISTRIBUTION: Known only from Cook County.

This species is similar to *J. alpinoarticulatus* and often confused with it. *Juncus articulatus* differs by having larger heads 8–10 mm across and long-apiculate capsules. On rare occasions this species may creep along the ground, rooting at the nodes. It flowers from July to October.

Juncus debilis Gray, Man. Bot. N. U.S. 506. 1848. Fig. A15.
Juncus acuminatus Michx. var. *debilis* (Gray) Engelm. Trans. Acad. Sci. St. Louis 2:463. 1868.

Cespitose perennial from short rhizomes; stems slender, erect or spreading when in water, to 40 cm long; leaves terete, septate, 0.5–1.0 mm wide, the lowest leaf sometimes reduced to a sheath; involucral leaf shorter than the inflorescence; inflorescence diffuse, with several widely divergent rays; heads 3–35, 3–5 mm wide, wedge-shaped, 2- to 5- (10-) flowered; perianth segments linear-lanceolate, acuminate; sepals 1.2–2.8 mm long, about as long as the petals; stamens 3; capsules narrowly ovoid to ellipsoid, acute, longer than the perianth, pale cinnamon-brown; seeds narrowly ellipsoid, 0.3–0.4 mm long, apiculate.

COMMON NAME: Weak Rush.

HABITAT: In streams.

RANGE: New York to Missouri, south to eastern Texas and northern Florida.

A14. Juncus articulatus (Jointed Rush). a. Habit. b. Cluster of capsules. c. Open flower. d. Capsule.

A15. Juncus debilis (Weak Rush). a. Habit. b. Auricles at top of sheath. c. Capsule with perianth. d. Seed.

ILLINOIS DISTRIBUTION: Rare in southwestern Illinois: Jackson County.

This species has a diffuse inflorescence with up to 30 small clusters of flowers per plant. It differs from the similar appearing *J. alpinoarticulatus* and *J. articulatus* by the lack of an apiculus on the seed, by its pale cinnamon-brown seeds, and by its very narrow leaves. It flowers from May to August.

Juncus subcaudatus (Engelm.) Coville & S. F. Blake, Proc. Bot. Soc. Wash. 31:45. 1918. Fig. A16.

Juncus canadensis J. Gay var. *subcaudatus* Engelm. Trans. Acad. Sci. St. Louis 2:474. 1868.

Cespitose perennial with few short rhizomes; stems to 80 cm tall; leaves terete, septate, up to 15 cm long, 1–2 mm wide, the lowest leaf sometimes reduced to a sheath; involucral bract shorter than the inflorescence; inflorescence with spreading branches, to 15 cm long; heads 6–35, hemispherical, 5–10 mm in diameter, 5- to 10-flowered; perianth segments lanceolate, acuminate; sepals 2.0–3.2 mm long, slightly shorter than the petals; stamens 3; capsule narrowly ovoid to ellipsoid, obtuse to acute, short-beaked, slightly longer than the perianth segments; seeds 0.7–1.2 mm long, with tail-like appendages at each end.

COMMON NAME: Short-tailed Rush.

HABITAT: Along streams.

RANGE: Newfoundland to southeastern Missouri and western Kentucky, east to North Carolina.

ILLINOIS DISTRIBUTION: Jackson County: Oakwood Bottoms, near Gorham, collected by Dan K. Evans.

Because of its small hemispherical heads, this species resembles *J. acuminatus*. It differs from *J. acuminatus* by its larger seeds and fewer flowers per head. This species flowers from July to October.

Juncus validus Coville, Bull. Torrey Club 22:305. 1895. Fig. A17.

Cespitose perennial from short rhizomes; stems stout, up to 1 m tall, hollow; leaves septate, more or less flat but tending to be a little terete and hollow, up to 45 cm long, 3–7 mm wide; involucral bract shorter than the inflorescence; inflorescence compact to open and branched, to 15 cm long; heads 15–75, mostly spherical, up to 10 mm in diameter, 25- to 70-flowered; perianth segments linear-lanceolate, acuminate; sepals 3.5–4.5 mm long, slightly longer than the petals; stamens 3; capsule lanceoloid, long-beaked, longer than the perianth segments; seeds 0.3–0.4 mm long, apiculate at both ends.

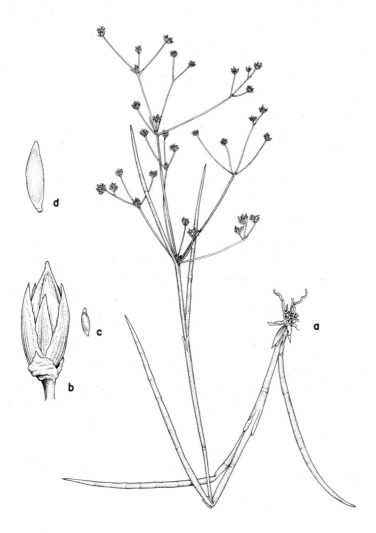

A16. Juncus subcaudatus (Short-tailed Rush). a. Habit. b. Capsule with perianth. c. Seed. d. Seed, enlarged.

A17. Juncus validus (Stout Rush). a. Habit. b. Capsule with perianth.

COMMON NAME: Stout Rush.

HABITAT: Edge of lake.

RANGE: North Carolina to southern Missouri, south to Texas and Florida.

ILLINOIS DISTRIBUTION: Known only from Horseshoe Lake, Alexander County.

This exceptionally stout species has leaves that are nearly flat but are still hollow and septate. It is the only *Juncus* with that type of leaf. This species flowers from July to September.

SUMMARY OF THE TAXA TREATED IN THIS VOLUME

Orders and Families	Genera	Species	Varieties or Subspecies
Order 1. **Alismales**			
Family 1. Butomaceae	1	1	
Family 2. Alismaceae	3	15	
Family 3. Hydrocharitaceae	3	5	
Order 2. **Zosterales**			
Family 4. Juncaginaceae	1	2	
Family 5. Scheuchzeriaceae	1	1	
Family 6. Potamogetonaceae	2	24	
Family 7. Ruppiaceae	1	1	
Family 8. Zannichelliaceae	1	1	
Order 3. **Najadales**			
Family 9. Najadaceae	1	5	1
Order 4. **Arales**			
Family 10. Acoraceae	1	2	
Family 11. Araceae	5	6	1
Family 12. Lemnaceae	4	17	
Order 5. **Typhales**			
Family 13. Sparganiaceae	1	5	
Family 14. Typhaceae	1	3	
Order 6. **Commelinales**			
Family 15. Xyridaceae	1	2	
Family 16. Commelinaceae	2	8	1
Family 17. Pontederiaceae	4	7	
Family 18. Juncaceae	2	31	3
Totals	35	136	6

GLOSSARY

Acaulescent. Seemingly without aerial stems.

Achene. A type of one-seeded, dry, indehiscent fruit with the seed coat not attached to the mature ovary wall.

Acicular. Needle-like.

Actinomorphic. Having radial symmetry; regular, in reference to a flower.

Acuminate. Gradually tapering to a point.

Adnate. Fusion of dissimilar parts.

Alternate. Referring to the condition of structures arising singly along an axis; opposed to opposite.

Annulus. A ring-like structure in the flower of *Thismia*.

Anther. The terminal part of a stamen which bears pollen.

Anthesis. Flowering time.

Antrorse. Projecting forward.

Apically. At the apex.

Apiculate. Abruptly short-pointed at the tip.

Apocarpy. A condition in the flower where more than one free pistil occurs.

Appressed. Lying flat against the surface.

Arborescent. Becoming tree-like.

Arching. Moderately curving.

Areola (pl., **areolae**). A small area between leaf veins.

Aristate. Bearing an awn.

Aristulate. Short-awned.

Attenuate. Gradually becoming narrowed.

Auriculate. Bearing an ear-like process.

Axil. The angle between the base of a structure and the axis from which it arises.

Axile. On the axis, referring to the place of attachment of the ovules.

Axillary. Borne from an axil.

Basal. Confined to the lowest part.

Beak. A terminal projection.

Berry. A type of fruit where the seeds are surrounded only by fleshy material.

Bicostate. Having two veins or ribs.

Biglandular. Bearing two glands.

Bilobed. Bearing two lobes.

Biseriate. In two rows or series.

Bisexual. Referring to a flower which contains both stamens and pistils.

Blade. The green, flat, expanded part of the leaf.

Bract. An accessory structure at the base of many flowers, usually appearing leaflike.

Bracteole. A secondary bract.

Bristle. A stiff hair or hairlike growth; a seta.

Bulb. An underground, vertical stem with both scaly and fleshy leaves.

Bulblet. A small bulb.

Bulbous. Bearing a swollen base.

Callosity. Any hardened thickening.

Calyx. The outermost ring of structures of a flower, composed of sepals.

Campanulate. Bell-shaped.

Capillary. Thread-like.

Capitate. Forming a head.

Capsule. A dry, dehiscent fruit composed of more than one carpel.

Carpel. A simple pistil, or one member of a compound pistil.

Cartilaginous. Firm but flexible.

Caruncle. A fleshy outgrowth near the point of attachment of a seed.

Carunculate. Bearing a fleshy outgrowth near its point of attachment.

Caudate. With a tail-like appendage.

Caudex. (pl., **caudices**). The woody base of a perennial plant.

Caulescent. Having an aerial stem.

Cauline. Belonging to a stem.

Cavernous. Hollowed out.

Cespitose. Growing in tufts.

Chartaceous. Papery.

Cilia. Marginal hairs.

Ciliate. Bearing cilia.

Ciliolate. Bearing small cilia.

Clasping. Referring to a leaf whose base encircles the stem.

Claw. A narrow, basal stalk, particularly of a petal.

Compressed. Flattened.

Concave. Curved on the inner surface; opposed to convex.

Connate. Union of like parts.

Connective. That portion of the stamen between the two anther halves.

Connivent. Coming in contact; converging.

Contiguous. Adjoining.

Convex. Rounded on the outer surface; opposite of concave.

Convolute. Rolled lengthwise.

Coralline. Having the texture of coral.

Coralloid. Resembling coral.

Cordate. Heart-shaped.

Coriaceous. Leathery.

Corm. An underground, vertical stem with scaly leaves, differing from a bulb by lacking fleshy leaves.

Corolla. The ring of structures of a flower just within the calyx, composed of petals.

Corona. A crown of petal-like structures, as in *Narcissus*.

Corymb. A type of inflorescence where the pedicellate flowers are arranged along an elongated axis but with the flowers all attaining about the same height.

Creeping. Spreading on the surface of the ground.

Crested. Bearing a ridge.

Crisped. Curled.

Cross-striae. Markings perpendicular to the longitudinal axis.

Cruciform. Cross-shaped.

Cucullate. Hood-shaped.

Cuneate. Wedge-shaped or tapering at the base.

Cupular. Shaped like a small cup.

Cuspidate. Terminating in a very short point.

Cyme. A type of broad and flattened inflorescence in which the central flowers bloom first.

Cymose. Bearing a cyme.

Deciduous. Falling away.
Decumbent. Lying flat, but with the tip ascending.
Deflexed. Turned downward.
Dehiscent. Splitting at maturity.
Deltoid. Triangular.
Dentate. With sharp teeth, the tips of which project outward.
Denticulate. With small, sharp teeth, the tips of which project outward.
Dilated. Swollen; expanded.
Dimorphic. Having two forms.
Dioecious. With staminate flowers on one plant, pistillate flowers on another.
Distal. Remote from the point of attachment.
Distended. Swollen.
Divaricate. Spreading.
Dorsal. That surface turned away from the axis; abaxial.
Drupaceous. Drupe-like.
Drupe. A type of fruit in which the seed is surrounded by a hard, dry covering which, in turn, is surrounded by fleshy material.

Eglandular. Without glands.
Ellipsoid. Referring to a solid object which is broadest at the middle, gradually tapering to both ends.
Elliptic. Broadest at middle, gradually tapering equally to both ends.
Emersed. Rising above the surface of the water.
Emucronulate. Without a short, abrupt tip.
Endosperm. Food-storage tissue found outside the embryo.

Ensiform. Sword-shaped.
Ephemeral. Lasting only a short time.
Epicarp. The outermost layer of the fruit.
Epunctate. Without dots.
Erose. With an irregularly notched margin.
Exudate. Secreted material.

Facial. Referring to the front surface.
Falcate. Sickle-shaped.
Fascicle. A cluster; a bundle.
Fibrous. Referring to roots borne in tufts.
Filament. That part of the stamen supporting the anther.
Filiform. Threadlike.
Flabellate. Fan-shaped.
Flexuous. Zigzag.
Foliaceous. Leaf-like.
Follicle. A type of dry, dehiscent fruit which splits along one side at maturity.
Friable. Breaking easily into small particles.
Frond. In this volume, the vegetative structure in the Lemnaceae.
Funnelform. Shaped like a funnel.
Fusiform. Spindle-shaped; tapering at both ends.

Galea. A hooded portion of a perianth.
Glabrous. Without pubescence or hairs.
Glaucescent. Becoming covered with a whitish bloom which can be rubbed off.
Glaucous. With a whitish covering which can be rubbed off.
Globoid. Referring to a solid body which is round.

Globose. Round; globular.

Glomerulate. Forming small heads.

Glumaceous. Resembling a scale.

Glutinous. Covered with a sticky secretion.

Hastate. Spear-shaped; said of a leaf which is triangular with spreading basal lobes.

Head. A type of inflorescence in which several sessile flowers are clustered together at the tip of a peduncle.

Hemispherical. Half-spherical.

Herbaceous. Not woody; dying back all the way to the ground in winter.

Hood. That part of an orchid flower which is strongly concave and arching.

Hyaline. Transparent.

Indehiscent. Not splitting open at maturity.

Inferior. Referring to the position of the ovary when it is surrounded by the adnate portion of the floral tube or is embedded in the receptacle.

Inflorescence. A cluster of flowers.

Internode. The area between two adjacent nodes.

Involucral. Referring to a circle of bracts which subtend a flower cluster.

Involute. Rolled inward.

Irregular. In reference to a flower having no symmetry at all.

Keeled. Possessing a ridgelike process.

Lanceolate. Lance-shaped; broadest near base, gradually tapering to the narrow apex.

Lanceoloid. Referring to a solid object which is broadest near base, gradually tapering to the narrow apex.

Linear. Narrow and approximately the same width at either end and the middle.

Lip. The lowermost, often greatly modified petal, in the flower of an orchid.

Lobe. A projection separated from each adjacent projection by a sinus.

Locular. Referring to the cells of a compound ovary.

Locule. A cell or cavity of a compound ovary.

Loculicidal. Said of a capsule which splits down the dorsal suture of each cell.

Lodicule. A membranous scale found within some grass flowers, possibly representing the perianth.

Lustrous. Shiny.

Marginate. With a definite margin.

Mealy. Having a granular texture.

Median. Pertaining to the middle.

Membranous. Like a membrane; thin.

Moniliform. Constricted at regular intervals to resemble a string of beads.

Monoecious. Bearing both sexes in separate flowers on the same plant.

Mucro. A short abrupt tip.

Mucronate. Said of a leaf with a short, terminal point.

Mucronulate. Said of a leaf with a very short, terminal point.

Net-veined. Having veins forming closed meshes.

Node. That place on the stem from which leaves and branchlets arise.

Nodose. Knotty.

Nutlet. A small nut.

Oblanceolate. Reverse lance-shaped; broadest at apex, gradually tapering to narrow base.

Oblique. One-sided; asymmetrical.

Oblong. Broadest at the middle, and tapering to both ends, but broader than elliptic.

Oblongoid. Referring to a solid object which, in side view, is nearly the same width throughout, but broader than linear.

Obovoid. Referring to a solid object which is broadly rounded at the apex, becoming narrowed below.

Obpyramidal. Referring to an upside-down pyramid.

Obtuse. Rounded at the apex.

Opaque. Incapable of being seen through.

Opposite. Referring to the condition of two like structures arising from the same point and across from each other.

Orbicular. Round.

Orthotropous. Referring to an ovule which is borne upright.

Ovary. The lower swollen part of the pistil which produces the ovules.

Ovate. Broadly rounded at base, becoming narrowed above; broader than lanceolate.

Ovoid. Referring to a solid object which is broadly rounded at the base, becoming narrowed above.

Ovule. The egg-producing structure found within the ovary.

Panduriform. Fiddle-shaped.

Panicle. A type of inflorescence composed of several racemes.

Papillose. Bearing pimple-like processes.

Papule. A pimple-like projection.

Parallel-veined. Having veins running in the same direction and not meeting.

Pedicel. The stalk of a flower of an inflorescence.

Peduncle. The stalk of an inflorescence.

Pedunculate. Bearing a peduncle.

Pellucid. Being transparent, in reference to spots or dots.

Peltate. Attached away from the margin, in reference to a leaf.

Pendent. Suspended; overhanging.

Pendulous. Hanging.

Perennial. Living more than two years.

Perfect. Bearing both stamens and pistils in the same flower.

Perfoliate. Referring to a leaf which appears to have the stem pass through it.

Perianth. Those parts of a flower including both the calyx and corolla.

Persistent. Remaining attached.

Petal. One segment of the corolla.

Petaloid. Resembling a petal in texture and appearance.

Petiolate. Bearing a petiole, or leafstalk.

Petiole. The stalk of a leaf.

Phyllodia. Dilated petioles modified to resemble and function as leaves.

Pilose. Bearing soft hairs.

Pilosulous. Bearing short, soft hairs.

Pistil. The ovule-producing organ of a flower normally composed of an ovary, a style, and a stigma.

Pistillate. Bearing pistils but not stamens.

Placentation. Referring to the manner in which the ovules are attached in the ovary.

Plicate. Folded.

Podogyne. A stalk in *Ruppia* on top at which the fruit is produced.

Prophyll. A bracteole.

Prophyllate. Bearing a bracteole at the base of a flower.

Prostrate. Lying flat.

Puberulent. With minute hairs.

Pubescent. Bearing some kind of hairs.

Punctation. A dot or dots.

Quadrangular. Four-angled.

Quadrate. Four-sided.

Raceme. A type of inflorescence where pedicellate flowers are arranged along an elongated axis.

Ranked. Referring to the number of planes in which structures are borne.

Receptacle. That part of the flower to which the perianth, stamens, and pistils are usually attached.

Reflexed. Turned downward.

Regular. Having radial symmetry (actinomorphic) or bilateral symmetry (zygomorphic).

Reniform. Kidney-shaped.

Resin. A usually sticky secretion found in various parts of certain plants.

Resupinate. Upside down.

Reticulate. Resembling a network.

Retuse. Shallowly notched at a rounded apex.

Rhizome. An underground horizontal stem, bearing nodes, buds, and roots.

Rhombic. Becoming quadrangular.

Ribbed. Nerved; veined.

Root cap. A group of cells borne externally at the tip of a root.

Rosette. A cluster of leaves in a circular arrangement at the base of a plant.

Rotate. Flat and circular.

Rugose. Wrinkled.

Rugulose. With small wrinkles.

Saccate. Sac-shaped.

Sagittate. Shaped like an arrowhead.

Salverform. Referring to a tubular corolla which abruptly expands into a flat limb.

Saprophyte. A type of plant living on dead or decaying organic matter, usually through the medium of mycorrhizae.

Scaberulous. Minutely roughened; slightly rough to the touch.

Scabrous. Rough to the touch.

Scale. A minute epidermal outgrowth, sometimes becoming green and replacing the leaf in function.

Scape. A leafless stalk bearing a flower or inflorescence.

Scapose. Possessing a leafless flowering stem.

Scarious. Thin and membranous.

Scurfy. Bearing scaly particles.

Secund. Borne on one side.

Sepal. One segment of the calyx.

Sepaloid. Appearing like a sepal.

Septate. With cross walls.

Septicidal. Said of a capsule which splits between the locules.

Serrulate. With very small teeth, the tips of which project forward.

Sessile. Without a stalk.

Setaceous. Bearing bristles, or setae.

Setose. Bearing bristles.

Sheath. A protective covering.

Shoot. The developing stem with its leaves.

Spadix. A fleshy axis in which flowers are embedded.

Spathe. A large sheathing bract subtending or usually enclosing an inflorescence.

Spatulate. Oblong, but with the basal end elongated.

Spike. A type of inflorescence where sessile flowers are arranged along an elongated axis.

Spinescent. Becoming spiny.

Spinule. A small spine.

Spinulose. With small spines.

Stamen. The pollen-producing organ of a flower composed of a filament and an anther.

Staminate. Bearing stamens but not pistils.

Staminodium (pl., staminodia). A sterile stamen.

Stipitate. Bearing a stipe or stalk.

Stipule. A leaf-like or scaly structure found at the point of attachment of a leaf to the stem.

Stoloniferous. Bearing runners or slender horizontal stems on the surface of the ground.

Stramineous. Straw-colored.

Style. That part of the pistil between the ovary and the stigma.

Subacute. Nearly acute.

Subcapitate. Nearly head-like.

Subligneous. Nearly woody.

Submersed. Covered with water.

Suborbiculate. Nearly round.

Subsessile. Nearly without a stalk.

Subulate. Drawn to an abrupt sharp point.

Subuloid. Referring to a solid object which is drawn to an abrupt short point.

Suffused. Spread throughout; flushed.

Superior. Referring to the position of the ovary when the free floral parts arise below the ovary.

Supra-axillary. Borne above the axil.

Tendril. A spiralling coiling structure which enables a climbing plant to attach itself to a supporting body.

Tenuous. Slender.

Terete. Rounded in cross-section.

Thalloid. Possessing an undifferentiated plant body, that is, without roots, stems or leaves.

Throat. The opening at the apex of a corolla tube where the limb arises.

Translucent. Partly transparent.

Trilocular. With three cavities.

Trimorphic. Having three forms.

Truncate. Abruptly cut across.

Tuber. An underground fleshy stem formed as a storage organ at the end of a rhizome.

Tubercle. A small wart-like process.

Tubular. Shaped like a tube.

Tunicated. Covered with concentric coats.

Turbinate. Shaped like a top.

Turgid. Tightly inflated.

Umbel. A type of inflorescence in which the flower stalks arise from the same level.

Umbonate. With a stout projection at the center.

Unarmed. Without prickles or spines.

Undulate. Wavy.

Unisexual. Bearing either stamens or pistils in one flower.

Utricle. A small one-seeded, indehiscent fruit with a thin covering.

Valve. That part of a capsule which splits.

Venation. The pattern of the veins.

Ventral. That surface turned toward the axis; adaxial.

Villous. With long soft erect hairs.

Viscid. Sticky.

Whorl. An arrangement of three or more structures at a point on the stem.

Whorled. Referring to the condition of three or more like structures arising from the same point.

Winged. Bearing a flat lateral outgrowth.

Zygomorphic. Bilaterally symmetrical.

LITERATURE CITED

Anderson, E. and R. E. Woodson, Jr. 1935. The species of *Tradescantia* indigenous to the United States. Contributions from the Arnold Arboretum 9:1–132.

Beal, E. O. 1964. *Sagittaria* in A. E. Radford, H. E. Ahles, and C. R. Bell, Guide to the vascular flora of the Carolinas, The Book Exchange, University of North Carolina, Chapel Hill, N. C., pp. 54–56.

Blake, S. F. 1912. The forms of *Peltandra virginica*. Rhodora 14: 102–6.

Bogin, C. 1955. Revision of the genus *Sagittaria* (Alismataceae). Memoirs of the New York Botanical Garden 9(2):179–233.

Bowden, W. M. 1940. Diploidy, polyploidy, and winter hardiness relationships in the flowering plants. American Journal of Botany 27:357–71.

———. 1945. A list of chromosome numbers in higher plants. II. Menispermaceae to Verbenaceae. American Journal of Botany 32:191–201.

Brendel, F. 1859. Additions and annotations to Mr. Lapham's catalogue of Illinois plants. Transactions of the Illinois State Agricultural Society 3:583–85.

———. 1887. Flora Peoriana. Peoria, Illinois.

Brown, W. V. 1947. Cytological studies in the Alismaceae. Botanical Gazette 108:262–67.

Buchenau, F. 1906. Juncaceae. In Engler, Das Pflanzenreich 25 (IV. 36):1–284.

Chase, S. S. 1947. Polyploidy in an immersed aquatic angiosperm. American Journal of Botany 34:581–82.

Coville, F. V. 1893. *Juncus marginatus* and its varieties. Proceedings of the Biological Society of Washington 8:121–28.

———. 1895. *Juncus scirpoides* and its immediate relatives. Bulletin of the Torrey Botanical Club 22:302–5.

Cronquist, A. 1968. The evolution and classification of flowering plants. Houghton Mifflin Company, Boston, 396 pp.

Curtis, J. 1862. British entomology. Volume VII. Lovell Reeve and Company, London.

Dark, S. O. S. 1932. Meiosis in diploid and triploid *Hemerocallis*. New Phytologist 31(5):310–20.

Darlington, C. D. and C. J. Vosa. 1963. Bias in the internal coiling direction of chromosomes. Chromosoma 13:609–22.

Darlington, C. D. and A. P. Wylie. 1956. Chromosome atlas of flowering plants. The Macmillan Company, New York.

Daubs, E. H. 1965. A monograph of Lemnaceae. Illinois Biological Monographs No. 34, The University of Illinois Press, Urbana. 118 pp.

DeFilipps, R. A. 1964. A taxonomic study of *Juncus* in Illinois. The American Midland Naturalist 71:296–319.

————. 1966. Distribution of *Juncus* in Illinois. Phytologia 13: 44–64.

Dobbs, R. J. 1963. Flora of Henry County, Illinois. Natural Land Institute, Rockford, Illinois. 350 pp.

Ellis, J. E. 1963. The genus *Potamogeton* in Illinois. Unpublished Master's thesis, Southern Illinois University.

Engelmann, G. A. 1866, 1868. A revision of the North American species of *Juncus*. Transactions of the Academy of St. Louis 2:424–58, 459–98.

Fassett, N. C. 1955. *Echinodorus* in the American tropics. Rhodora 57:133–56, 174–88, 202–12.

Fernald, M. L. 1932. The linear-leaved North American species of *Potamogeton*, Section Axillares. Memoirs of the American Academy of Arts and Sciences 17(1):1–183.

————. 1940. The varieties of *Commelina erecta*. Rhodora 42: 435–41.

————. 1945. Botanical specialties of the Seward Forest and adjacent areas of southeastern Virginia. Rhodora 47:93–142.

————. 1946. The North American representatives of *Alisma Plantago-aquatica*. Rhodora 48:86–88.

————. 1950. Gray's manual of botany. Edition 8. The American Book Company, New York. 1632 pp.

Fernald, M. L. and K. M. Wiegand. 1914. The genus *Ruppia* in eastern North America. Rhodora 16:119–27.

————. 1923. Notes on some plants of the Ontario and St. Lawrence basins, New York. Rhodora 25:205–14.

Fore, P. L. and R. H. Mohlenbrock. 1966. Two new naiads from Illinois and distributional records of the Naiadaceae. Rhodora 68: 216–20.

Gadella, T. W. J. and K. Kliphuis. 1963. Chromosome numbers of flowering plants in the Netherlands. Acta Bot. Neerl. 12:195–230.

Gleason, H. A. 1952. The new Britton and Brown illustrated flora of the northeastern United States. Volume 1. The New York Botanical Garden, New York.

———— and A. Cronquist. 1963. Manual of vascular plants of northeastern United States and adjacent Canada. D. Van Nostrand and Company, Princeton, Toronto, New York, and London. 810 pp.

Graebner, P. 1907. Potamogetonaceae. In Engler, Das Pflanzenreich 29:115. 1907.

Hagström, J. O. 1916. Critical researches on the Potamogetons. Kungl. Svenska Vetenskapsaked. Handl. 55(5).

Harada, I. 1942. Chromosome studies on *Potamogeton*. Japanese Journal of Genetics 18:92.

Hegelmaier, F. 1868. Die Lemnaceen, eine monographische untersuchung. Leipzig. 160 pp.

————. 1896. Systematische uebersicht der Lemnaceen. Botanische Jahrbücher 21:268–305.

Heiser, C. B. and T. W. Whitaker. 1948. Chromosome number, polyploidy, and growth habit in California weeds. American Journal of Botany 35:179–86.

Hendricks, A. J. 1957. A revision of the genus *Alisma*. The American Midland Naturalist 58:470–93.

Hermann, F. J. 1940. Juncaceae, in C. C. Deam, Flora of Indiana. Department of Conservation, Division of Forestry, Indianapolis, Indiana.

Hicks, L. E. 1937. The Lemnaceae of Indiana. The American Midland Naturalist 18:774–89.

Hill, E. J. 1896. Notes on the flora of Chicago and vicinity. II. Botanical Gazette 21:118–22.

Huttleston, D. G. 1949. The three subspecies of *Arisaema triphyllum*. Bulletin of the Torrey Botanical Club 76:407–13.

Itô, T. 1942. Chromosomen und Sexualität von der Araceae I. Somatische Chromosomenzahlen einiger arten. Cytologia 12: 313–25.

Jones, G. N. 1945. Flora of Illinois; a reply to M. L. Fernald. The American Midland Naturalist 34:271–84.

————. 1963. Flora of Illinois. Third edition. University of Notre Dame Press, South Bend, Indiana.

————, G. D. Fuller, G. S. Winterringer, H. E. Ahles, and A. A. Flynn. 1955. Vascular plants of Illinois. The University of Illinois Press, Urbana, and The Illinois State Museum, Springfield (Museum Scientific Series, Volume VI). 593 pp.

Kibbe, A. 1952. A botanical study and survey of a typical midwestern county (Hancock County, Illinois). Carthage, Illinois. 425 pp.

Kral, R. 1960. The genus *Xyris* in Florida. Rhodora 62:295–319.

Lapham, I. A. 1857. Catalogue of the plants of the State of Illinois. Transactions of the Illinois State Agricultural Society 2:492–550.

Lohammar, G. 1931. The chromosome numbers of *Sagittaria natans* Pallas and *S. sagittifolia* L. Svensk. Bot. Tidskr. 25:32–35.

Löve, A. and D. Löve. 1944. Cyto-taxonomical studies on boreal plants. III. Some new chromosome numbers of Scandinavian plants. Arkiv. Bot. 31A(12):1–22.

————. 1948. Chromosome numbers of northern plant species. Ingolfsprent, Reykjavik, Iceland.

Manton, I. 1949. Notes on the cytology of *Scheuchzeria* from Rannoch Moor. Watsonia 1:36.

Matsuura, H. and T. Sutô. 1935. Contributions to the idiogram study in phanerogamous plants. Journal of the Faculty of Science of Hokkaido University 5:33–75.

McClure, J. W. and R. E. Alston. 1966. A chemotaxonomic study of Lemnaceae. American Journal of Botany 53(9):849–60.

McDougall, W. B. 1936. Fieldbook of Illinois wild flowers. Illinois Natural History Survey, Urbana. 406 pp.

Mead, S. B. 1846. Catalogue of plants growing spontaneously in the State of Illinois, the principal part near Augusta, Hancock County. Prairie Farmer 6:35–36, 60, 93, 119–22.

Miki, S. 1937. The origin of *Najas* and *Potamogeton*. Botanical Magazine of Tokyo 51:290–480.

Mohlenbrock, R. H. 1962. On the occurrence of *Lilium superbum* in Illinois. Castanea 27:173–76.

———— and J. E. Ozment. 1967. Flowering plants new to Illinois. Transactions of the Illinois State Academy of Science 60(2): 186–88.

Mohlenbrock, R. H. and J. W. Richardson. 1967. Contributions to a flora of Illinois. No. 1. The Order Alismales. Transactions of the Illinois State Academy of Science 60(3):259–71.

Morong, T. 1893. The Naiadaceae of North America. Memoirs of the Torrey Botanical Club 3:1–65.

Muenscher, W. C. 1944. Aquatic plants of the United States. New York. 374 pp.

Ogden, E. C. 1943. The broad-leaved species of *Potamogeton* of North America north of Mexico. Rhodora 45:57–105, 119–63, 171–214.

Oleson, E. M. 1941. Chromosomes of some Alismaceae. Botanical Gazette 103:108–202.

Palmgren, O. 1939. Cytological studies in Potamogeton. Botaniska Notiser 1939:246–48.

Patterson, H. N. 1874. A list of plants collected in the vicinity of Oquawka, Henderson County. Oquawka, Illinois. 18 pp.

————. 1876. Catalogue of the phaenogamous and vascular cryptogamous plants of Illinois. Oquawka, Illinois. 54 pp.

Pepoon, H. S. 1909. The cliff flora of JoDaviess County. Transactions of the Illinois State Academy of Science 2:32–37.

————. 1927. An annotated flora of the Chicago area. Bulletin of the Chicago Academy of Science 8:1–554.

Pfeiffer, N. E. 1914. Morphology of *Thismia americana*. Botanical Gazette 57:122–35.

Pogan, E. 1963. Taxonomical value of *Alisma triviale* Pursh and *Alisma subcordatum* Rafin. Canadian Journal of Botany 41: 1011–13.

Robinson, B. L. 1903. The generic position of *Echinodorus parvulus*. Rhodora 5:85–89.

Schleiden, W. 1839. Prodromus monographie Lemnacearum. Linnaea 13:385–92.

Schneck, J. 1876. Catalogue of the flora of the Wabash Valley. Annual Report of the Geological Survey of Indiana 7:504–79.

Snogerup, S. 1963. Studies in the genus Juncus III. Observations on the diversity of chromosome numbers. Botaniska Notiser 116: 142–56.

Sokolovskaya, A. P. 1963. Geographical distribution of polyploidy in plants. Vest. Leningrad Univ. No. 15, Ser. Biol.:38–52.

Sorsa, V. 1963. Chromosomenzahlen Finnischer Kormophyton II. Ann. Acad. Sci. Fenn. Ser. A. IV. Biol. 68:1–14.

Stern, K. R. 1961. Chromosome numbers in nine taxa of *Potamogeton*. Bulletin of the Torrey Botanical Club 88:411–14.

St. John, H. 1965. Monograph of the genus *Elodea*: Part 4 and summary. Rhodora 67:1–35, 155–80.

Taylor, N. 1909. Zannichelliaceae, in North American Flora 17:part 1:13–27.

Thompson, C. H. 1898. A revision of the American Lemnaceae occurring north of Mexico. Annual Report of the Missouri Botanical Garden 9:21–42.

Thorne, R. F. 1963. Some problems and guiding principles of angiosperm phylogeny. The American Naturalist 97:287–305.

———. 1968. Synopsis of a putatively phylogenetic classification of the flowering plants. Aliso 6(4):57–66.

Uhl, N. W. 1947. Studies in the floral morphology and anatomy of certain members of the Helobiae. Unpublished Dissertation, Cornell University, Ithaca, New York.

Vasey, G. 1861. Additions to the flora of Illinois. Transactions of the Illinois Natural History Society 1:139–43.

Weik, K. L. and R. H. Mohlenbrock. 1968. Contributions to a flora of Illinois. No. 3. The Family Lemnaceae. Transactions of the Illinois State Academy of Science 61:382–99.

Wherry, E. T. 1942. The relationship of *Lilium michiganense*. Rhodora 44:453–56.

Whitaker, T. W. 1934. Chromosome constitution in certain Monocotyledons. Journal of the Arnold Arboretum 15:135–43.

Winterringer, G. S. 1966. Aquatic vascular plants new for Illinois. Rhodora 68:221–22.

——— and R. A. Evers. 1960. New records for Illinois vascular plants. Illinois State Museum, Scientific Papers Series, Volume XI. 136 pp.

INDEX OF PLANT NAMES

Names in roman type are accepted names, while those in italics are synonyms and are not considered valid. Page numbers in bold refer to pages that have illustrations.

Robert H. Mohlenbrock taught botany at Southern Illinois University Carbondale for thirty-four years, obtaining the title of Distinguished Professor. Since his retirement in 1990, he has served as senior scientist for Biotic Consultants Inc., teaching wetland identification classes around the country. Mohlenbrock has been named SIU Outstanding Scholar and has received the SIU Alumnus Teacher of the Year Award, the College of Science Outstanding Teacher Award, and the Meritorious Teacher of the Year Award from the Association of Southeastern Biologists. Since 1984, he has been a monthly columnist for *Natural History* magazine. He is the author of 54 books and 565 publications.